EVALUATION OF CHEMICAL EVENTS
at Army Chemical Agent Disposal Facilities

Committee on Evaluation of Chemical Events at Army Chemical
Agent Disposal Facilities

Board on Army Science and Technology

Division on Engineering and Physical Sciences

NATIONAL RESEARCH COUNCIL
OF THE NATIONAL ACADEMIES

THE NATIONAL ACADEMIES PRESS
Washington, D.C.
www.nap.edu

THE NATIONAL ACADEMIES PRESS • 500 Fifth Street, N.W. • Washington, DC 20001

NOTICE: The project that is the subject of this report was approved by the Governing Board of the National Research Council, whose members are drawn from the councils of the National Academy of Sciences, the National Academy of Engineering, and the Institute of Medicine. The members of the committee responsible for the report were chosen for their special competences and with regard for appropriate balance.

This study was supported by Contract No. DAAD 19-01-C-0051 between the National Academy of Sciences and the Department of Defense. Any opinions, findings, conclusions, or recommendations expressed in this publication are those of the author(s) and do not necessarily reflect the views of the organizations or agencies that provided support for the project.

International Standard Book Number 0-309-08629-9

Cover: Decontaminated chemical munitions and containers at Johnston Atoll Chemical Agent Disposal System. Photographs for composite image courtesy of Colin Drury.

Additional copies of this report are available from the National Academies Press, 500 Fifth Street, N.W., Lockbox 285, Washington, D.C. 20055; (800) 624-6242 or (202) 334-3313 (in the Washington metropolitan area); Internet, http://www.nap.edu

Copyright 2002 by the National Academy of Sciences. All rights reserved.

Printed in the United States of America

THE NATIONAL ACADEMIES
Advisers to the Nation on Science, Engineering, and Medicine

The **National Academy of Sciences** is a private, nonprofit, self-perpetuating society of distinguished scholars engaged in scientific and engineering research, dedicated to the furtherance of science and technology and to their use for the general welfare. Upon the authority of the charter granted to it by the Congress in 1863, the Academy has a mandate that requires it to advise the federal government on scientific and technical matters. Dr. Bruce M. Alberts is president of the National Academy of Sciences.

The **National Academy of Engineering** was established in 1964, under the charter of the National Academy of Sciences, as a parallel organization of outstanding engineers. It is autonomous in its administration and in the selection of its members, sharing with the National Academy of Sciences the responsibility for advising the federal government. The National Academy of Engineering also sponsors engineering programs aimed at meeting national needs, encourages education and research, and recognizes the superior achievements of engineers. Dr. Wm. A. Wulf is president of the National Academy of Engineering.

The **Institute of Medicine** was established in 1970 by the National Academy of Sciences to secure the services of eminent members of appropriate professions in the examination of policy matters pertaining to the health of the public. The Institute acts under the responsibility given to the National Academy of Sciences by its congressional charter to be an adviser to the federal government and, upon its own initiative, to identify issues of medical care, research, and education. Dr. Harvey V. Fineberg is president of the Institute of Medicine.

The **National Research Council** was organized by the National Academy of Sciences in 1916 to associate the broad community of science and technology with the Academy's purposes of furthering knowledge and advising the federal government. Functioning in accordance with general policies determined by the Academy, the Council has become the principal operating agency of both the National Academy of Sciences and the National Academy of Engineering in providing services to the government, the public, and the scientific and engineering communities. The Council is administered jointly by both Academies and the Institute of Medicine. Dr. Bruce M. Alberts and Dr. Wm. A. Wulf are chair and vice chair, respectively, of the National Research Council.

www.national-academies.org

COMMITTEE ON EVALUATION OF CHEMICAL EVENTS AT ARMY CHEMICAL AGENT DISPOSAL FACILITIES

CHARLES E. KOLB, *Chair,* Aerodyne Research, Inc., Billerica, Massachusetts
DENNIS C. BLEY, Buttonwood Consulting, Inc., Oakton, Virginia
COLIN G. DRURY, University of Buffalo, New York
JERRY FITZGERALD ENGLISH, Cooper, Rose and English LLP, Summit, New Jersey
J. ROBERT GIBSON, Consultant, Wilmington, Delaware
HANK C. JENKINS-SMITH, Texas A&M University, College Station
WALTER G. MAY, NAE, University of Illinois at Urbana-Champaign
GREGORY McRAE, Massachusetts Institute of Technology, Cambridge
IRVING F. MILLER, Consultant, Chicago, Illinois
DONALD W. MURPHY, NAE, Consultant, Davis, California
ALVIN H. MUSHKATEL, Arizona State University, Tempe
LEIGH SHORT, Consultant, Mount Pleasant, South Carolina
LEO WEITZMAN, Consultant, West Lafayette, Indiana

National Research Council Staff

NANCY T. SCHULTE, Study Director (from June 2002)
MARGARET N. NOVACK, Study Director (to June 2002)
WILLIAM E. CAMPBELL, Administrative Officer
JIM MYSKA, Research Associate
PAMELA A. LEWIS, Senior Project Assistant
SONNETT HOSSANAH, Senior Project Assistant
CARTER W. FORD, Senior Project Assistant

BOARD ON ARMY SCIENCE AND TECHNOLOGY

JOHN E. MILLER, *Chair*, Oracle Corporation, Reston, Virginia
GEORGE T. SINGLEY III, *Vice Chair*, Hicks and Associates, Inc., McLean, Virginia
ROBERT L. CATTOI, Rockwell International (retired), Dallas, Texas
RICHARD A. CONWAY, Union Carbide Corporation (retired), Charleston, West Virginia
GILBERT F. DECKER, Walt Disney Imagineering (retired), Glendale, California
ROBERT R. EVERETT, MITRE Corporation (retired), New Seabury, Massachusetts
PATRICK F. FLYNN, Cummins Engine Company, Inc. (retired), Columbus, Indiana
HENRY J. HATCH, Army Chief of Engineers (retired), Oakton, Virginia
EDWARD J. HAUG, University of Iowa, Iowa City
GERALD J. IAFRATE, North Carolina State University, Raleigh
MIRIAM E. JOHN, California Laboratory, Sandia National Laboratories, Livermore
DONALD R. KEITH, Cypress International (retired), Alexandria, Virginia
CLARENCE W. KITCHENS, IIT Research Institute, Alexandria, Virginia
SHIRLEY A. LIEBMAN, CECON Group (retired), Holtwood, Pennsylvania
KATHRYN V. LOGAN, Georgia Institute of Technology (professor emerita), Roswell
STEPHEN C. LUBARD, S-L Technology, Woodland Hills, California
JOHN W. LYONS, U.S. Army Research Laboratory (retired), Ellicott City, Maryland
JOHN H. MOXLEY, Korn/Ferry International, Los Angeles, California
STEWART D. PERSONICK, Drexel University, Philadelphia, Pennsylvania
MILLARD F. ROSE, Radiance Technologies, Huntsville, Alabama
JOSEPH J. VERVIER, ENSCO, Inc., Melbourne, Florida

Staff

BRUCE A. BRAUN, Director
MICHAEL A. CLARKE, Associate Director
WILLIAM E. CAMPBELL, Administrative Officer
CHRIS JONES, Financial Associate
DANIEL E.J. TALMAGE, JR., Research Associate
DEANNA P. SPARGER, Senior Project Assistant

Preface

For over half a century the United States has maintained a stockpile of chemical weapons at Army depots distributed around the country. These weapons are now obsolete, and some have deteriorated to an alarming extent. Since 1990, in response to P.L. 99-145 and, later, P.L. 102-484, the Army's Program Manager for Chemical Demilitarization (PMCD) has been engaged in active destruction of the chemical weapons stockpile. Operation of the two initial chemical agent demilitarization facilities utilizing incinerator technology—Johnston Atoll Chemical Agent Disposal System (JACADS) and Tooele Chemical Agent Disposal Facility (TOCDF) (see Appendix A)—has achieved destruction of more than 23 percent of the original chemical agent tonnage (U.S. Army, 2001a) but has not been without incident. A number of chemical events have resulted in various levels of chemical agent migrating at higher than anticipated levels into areas within the plants themselves, and in a few incidents small amounts of chemical agent have been released into the ambient atmosphere (see Appendix B). Although none of these incidents resulted in agent releases large enough to be measured at the chemical demilitarization plant perimeters (U.S. Army, 2001c) and thus posed no threat to nearby communities, they did raise concern among affected public officials and citizens about the fundamental safety of incineration-based chemical demilitarization facilities, particularly the three third-generation incineration facilities scheduled to begin operation at depots near Anniston, Alabama; Umatilla, Oregon; and Pine Bluff, Arkansas.

STATEMENT OF TASK

This report was motivated by congressional concern that incidents at JACADS and TOCDF might indicate systemic safety issues with either the technology or the management and operational systems employed at those two initial chemical demilitarization facilities.

The Committee on Evaluation of Chemical Events at Army Chemical Agent Disposal Facilities, convened in April 2001 by the National Research Council (NRC), was charged with the following statement of task negotiated between the Army and the NRC:

> The National Research Council will assemble a committee to evaluate chemical events that have occurred at the Johnston Atoll Chemical Agent Disposal System (JACADS) and the Tooele Chemical Agent Disposal Facility (TOCDF). The committee will:
>
> • review process technology, operational activities (including training, operations and maintenance), and management by both the Army and its contractors to identify the causes of chemical events
>
> • review applicable risk management and safety programs
>
> • review emergency response activities that have occurred as a result of each chemical event, including information dissemination
>
> • review actions and changes that have occurred in response to each chemical event and evaluate the impact and adequacy of these actions and changes
>
> • visit JACADS and TOCDF to review facility configurations and to meet with personnel involved with operational activities, facility management, and emergency response
>
> • make recommendations regarding improvements in operational activities, facility management, and emergency response
>
> • review and recommend the needs to enable credible and more rapid investigation and corrective actions in response to future chemical events at chemical demilitarization sites, including consideration of needs of external stakeholders (e.g., regulators and concerned public).

To ensure that new facilities for the destruction of chemical agent are operated as safely as possible, the NRC was further asked to recommend how lessons learned from the events at JACADS and TOCDF should influence future operations, particularly at the new facilities in Alabama, Oregon, and Arkansas scheduled for completion and initial operations in the near future.

COMMITTEE COMPOSITION AND PROCESS

Committee members brought to their task extensive experience in chemical process engineering, chemical plant operations, human factors and ergonomics, industrial engineering, risk assessment and management, atmospheric sciences, environmental chemistry, toxicology, environmental regulations and law, emergency management, and public involvement and community relations (see Appendix H). In conducting this study, committee members drew on insights gained from their experiences in academia, chemical and related industries, federal and state agencies, private sector laboratories and consulting firms, and a law firm.

The committee first met as a whole in Washington, D.C., in May 2001 to hear Army briefings on JACADS and TOCDF general operations and chemical events. (Appendix I lists the committee's several meetings.) In early June many committee members attended an informational meeting on Capitol Hill hosted by Congressman Bob Riley (R-Ala.), who represents the region around the Anniston Chemical Demilitarization Facility, which is currently undergoing systemization and preoperational testing. Local government officials, emergency management professionals, and concerned citizens from the area near Anniston, Alabama, shared their perspectives with the committee. Committee members and staff also visited PMCD and its supporting contractors located at the Aberdeen Proving Ground, Maryland.

The committee made site visits to JACADS in late June 2001 and to TOCDF in late July 2001 where it investigated the operational history, management procedures, and evaluations of and responses to chemical events at these facilities and discussed these issues with contractors and PMCD personnel at many levels. At a meeting at Woods Hole, Massachusetts, in October 2001 the committee completed the bulk of the data-gathering process as well as much of the initial draft of its report. The November 2001 meeting, in Washington, D.C., was dedicated to completing the initial report draft. A portion of the committee also visited Anniston, Alabama, in early December 2001 to inspect a completed third-generation incineration facility and a storage depot with an extensive nearby population base. As a part of the visit the committee visited the County Emergency Response Facility, met with County Commissioners, and participated in a public meeting. A draft report suitable for NRC preview editing was produced subsequent to the Anniston visit. A final committee meeting in January 2002 focused on reviewing this draft, including refining the report's findings and recommendations.

The committee consulted with and received input from many stakeholders, both principals and agents, including personnel assigned to the office of the PMCD and its support contractors; contractor and subcontractor personnel responsible for operating chemical demilitarization facilities; former employees of chemical demilitarization facilities; congressional, state, and local officials; members of state citizen advisory committees; members of citizen activist groups; and local citizens. (See Appendixes C, D, and I.)

The committee has also benefited from previous NRC reports on the chemical demilitarization program. Many of these reports were prepared by a standing NRC committee, the Committee on Review and Evaluation of the Army Chemical Stockpile Disposal Program (the Stockpile Committee), which evaluates aspects of the disposal program at the request of the Army. Several of the Stockpile Committee reports provided background for this committee's study.

In preparing, reviewing, printing, and distributing this report, the National Research Council (NRC) and this committee are acting as an expert *agent* for several principals, including the U.S. Congress; the Army, which contracted with the NRC to perform the study; and the U.S. public.

The committee's goals for this report were to respond, as thoroughly as feasible in the short time allotted, to the concerns stakeholders have expressed about past chemical events at JACADS and TOCDF, to determine the impact of these events on ongoing operations at TOCDF, and to assess the implications of these events for the safe and efficient operation of incineration-based chemical demilitarization facilities scheduled to begin operation at Anniston, Umatilla, and Pine Bluff.

The committee greatly appreciates the support and assistance of National Research Council staff members Bruce A. Braun, Margaret Novack, Nancy Schulte, Bill Campbell, Jim Myska, Sonnett Hossanah, Pamela Lewis, and Carter Ford in the production of this report.

NOTE: Following preparation of this report two chemical events, one at TOCDF on July 15, 2002, and one at JACADS on August 12, 2002, have taken place. Although these incidents occurred after the committee completed its analysis, they are similar in nature to events analyzed by the committee and reinforce the validity of the findings and the utility of the recommendations presented in this report.

Charles E. Kolb, *Chair*
Committee on Evaluation of
   Chemical Events at Army Chemical
   Agent Disposal Facilities

Acknowledgment of Reviewers

This report has been reviewed in draft form by individuals chosen for their diverse perspectives and technical expertise, in accordance with procedures approved by the National Research Council's Report Review Committee. The purpose of this independent review is to provide candid and critical comments that will assist the institution in making its published report as sound as possible and to ensure that the report meets institutional standards for objectivity, evidence, and responsiveness to the study charge. The review comments and draft manuscript remain confidential to protect the integrity of the deliberative process. We wish to thank the following individuals for their review of this report:

Richard J. Ayen, Waste Management, Inc. (retired)
Judith A. Bradbury, Battelle Patuxent River
Dennis R. Downs, Utah Department of Environmental Quality
Charles A. Eckert, Georgia Institute of Technology
Richard S. Magee, Carmagan Engineering

Lewis S. Nelson, New York City Poison Control Center
George W. Parshall, E.I. DuPont de Nemours & Co. (retired)
William R. Rhyne, Informatics Corporation, and
Palmer W. Taylor, University of California, San Diego.

Although the reviewers listed above have provided many constructive comments and suggestions, they were not asked to endorse the conclusions or recommendations, nor did they see the final draft of the report before its release. The review of this report was overseen by Royce W. Murray, University of North Carolina, Chapel Hill. Appointed by the National Research Council, he was responsible for making certain that an independent examination of this report was carried out in accordance with institutional procedures and that all review comments were carefully considered. Responsibility for the final content of this report rests entirely with the authoring committee and the institution.

Contents

EXECUTIVE SUMMARY 1

1 THE CHEMICAL DEMILITARIZATION CHALLENGE 7
   Stockpile Content, Disposal Deadline, and Disposal Technology, 7
   Chemical Events, 9
      Chemical Events Associated with Disposal, 9
      Chemical Events Occurring During Storage, 10
   Tools for Assessing Hazards in the Operation of Chemical Stockpile
        Disposal Facilities, 11
      Prospective Risk Analysis Tools, 11
        Health Risk Assessment, 11
        Systems Hazard Analysis, 11
        Quantitative Risk Assessment, 11
      Retrospective Analysis Tools, 12
        Monitoring Systems, 12
        Chemical Event Investigations, 12
      Putting It All Together, 12
      Monitoring Methods, 14
      Event Analysis and Significance, 14
   Chemical Demilitarization Institutional Issues, 14
      Trust and Institutional Arrangements, 14
      The Institutional Setting of Chemical Demilitarization, 15
   Report Roadmap, 15

2 CAUSAL FACTORS IN EVENTS AT CHEMICAL DEMILITARIZATION
   FACILITIES 17
   Definitions, 17
   Sources of Input and Selection of Events for In-depth Analysis, 17
      The PMCD Incident List, 18
      The Calhoun County Commissioners' List, 18
      The Chemical Weapons Working Group Incident List, 18
      Notice of Violation Reports, 21
   Analysis of Selected Chemical Events, 22
      Causal Factors, 22
      Causal Tree Analysis of Two Events, 24
      General Observations, 24
      Specific Observations, 25

3	**RESPONSES TO CHEMICAL EVENTS AT BASELINE CHEMICAL DEMILITARIZATION FACILITIES**	26
	Formal Event Reporting Protocols, 26	
	Actual On-Site Responses, 27	
	December 3-5, 2000, Event at JACADS, 27	
	May 8-9, 2000, Event at TOCDF, 28	
	Observations, 28	
	External and Regulatory Responses to Chemical Events, 29	
	Applicable Statutes, Regulations, and Guidelines, 29	
	Memorandum of Understanding Between Deseret Chemical Depot and Tooele County, 29	
	Levels of Investigation, 30	
	Modeling Potential Population Exposure, 30	
	Emergency Response: Preparedness, Plans, Notification, and Coordination at TOCDF, 32	
	Public Responses to Chemical Events, 35	
4	**IMPLICATIONS OF PAST CHEMICAL EVENTS FOR ONGOING AND FUTURE CHEMICAL DEMILITARIZATION ACTIVITIES**	37
	Risk and Management of Change Programs Already in Place, 37	
	Safety Programs, 39	
	Programmatic Lessons Learned Program, 40	
	PLL Program Database, 40	
	Resultant Changes, 42	
5	**PREPARING FOR POTENTIAL FUTURE CHEMICAL EVENTS AT BASELINE CHEMICAL DEMILITARIZATION FACILITIES**	44
	Summary of Chemical Events Analyses, 44	
	Chemical Event Response and Review by Management, 44	
	Building on the Results of Risk Assessment, 45	
	Building a Safety Culture, 46	
	Operational Changes, 46	
	Worker Education, Training, and Involvement, 47	
	Desired Principal-*Agent* Interactions, 47	
	Rapid and Safe Restart Requirements, 49	
	Restarts After Changeovers and Maintenance, 49	
	Restarts After a Chemical Event, 49	
6	**FINDINGS AND RECOMMENDATIONS**	51
	REFERENCES	57
	APPENDIXES	
A	Specific Design Features of fhe Tooele Chemical Agent Disposal Facility Baseline Incineration System	61
B	Chronicle of Chemical Events and Other Occurrences at TOCDF and JACADS as Identified by PMCD	71
C	List of Individual Incidents from the Chemical Weapons Working Group	79
D	List of Individual Incidents from Calhoun County Commission, Anniston, Alabama	90
E	Additional Information Concerning Risk	97
F	Causal Tree Analysis of December 3-5, 2000, Event at JACADS	103
G	Memorandum of Understanding Between Deseret Chemical Depot and Tooele County for Information Exchange	106
H	Biographical Sketches of Committee Members	114
I	Committee Meetings	117

Figures, Tables, and Boxes

FIGURES

1-1 Location and size (percentage of original stockpile) of eight continental U.S. storage sites, 8
3-1 Component parts of an integrated system for modeling the impact of release of chemical agents, 31
4-1 TOCDF recordable injury rate 12-month rolling average, August 1996 (the start of agent operations) through December 2001, 39

A-1 Layout of the TOCDF, 62
A-2 Rocket-handling system, 63
A-3 Bulk handling system, 64
A-4 Projectile-handling system, 65
A-5 Mine-handling system, 66
A-6 Deactivation furnace system, 67
A-7 Metal parts furnace, 67
A-8 Liquid incinerator, 68
A-9 Dunnage furnace, 68
A-10 Pollution abatement system, 69

E-1 Schematic illustration of risk elements at a chemical agent and munitions storage and destruction site, 98
E-2 Contributors to the average public fatality risk from continued storage at Deseret Chemical Depot, 99
E-3 Comparison of risks to the public during processing at Deseret Chemical Depot and the Tooele Chemical Agent Disposal Facility, 100
E-4 Contributors to the average public fatality risk from disposal operations at DCD and TOCDF, 100
E-5 Contributors to the average risk of fatality for disposal-related workers at DCD and TOCDF, 101

F-1 Causal tree for December 3-5, 2000, JACADS event, 104-105

TABLES

2-1 Events on the PMCD List That Were Examined by the Committee, 19
2-2 Events on the PMCD List That Were Chosen by the Committee for Detailed Analysis, 19

xiii

2-3   Committee's Classification of 69 Items Cited in Notice of Violation Reports, 22
2-4   Frequency of Causal Factors in the Seven Incidents Analyzed by the Committee, 23

4-1   Issues and Factors in Assessing the Value of Change Options, 38

BOXES

1-1   Details on Airborne Chemical Agent Monitoring Methods and Standards at Chemical Demilitarization Facilities, 13

2-1   December 3-5, 2000, JACADS Event, 20
2-2   May 8-9, 2000, TOCDF Event, 21
2-3   An Example of Negative Effects of Mind-set, 25
3-1   Previous Concerns About and Recommendations for Achieving Efficient CSEPP Operations, 33

4-1   Additional PLL Program Components, 41

5-1   Examples of Observations That the Committee Concluded Were Uninformed, 48

Acronyms and Abbreviations

ACAMS	automatic continuous air monitoring system
AEGL	Acute Exposure Guideline Level
AMC	Army Materiel Command
ASC	allowable stack concentration
CAC	Citizens Advisory Commission
CAMDS	Chemical Agent Munitions Disposal System
CDC	Centers for Disease Control and Prevention
Chem demil	chemical demilitarization
CSDP	U.S. Chemical Stockpile Disposal Program
CSEPP	Chemical Stockpile Emergency Preparedness Program
CWC	Chemical Weapons Convention
CWWG	Chemical Weapons Working Group
DAAMS	depot area air monitoring system
DCD	Deseret Chemical Depot
DEQ	(Utah) Department of Environment Quality
DFS	deactivation furnace system
DoD	Department of Defense
DSHW	(Utah) Division of Solid and Hazardous Waste
DWL	drinking water level
ECP	engineering change proposal
ECR	explosive containment room
EG&G	Edgerton, Germerhausen and Grier (a contracting company)
EMIS	Emergency Management Information System
EOC	emergency operations center
EPA	Environmental Protection Agency
FEMA	Federal Emergency Management Agency
FPD	flame photometric detector
GAO	General Accounting Office
GB	sarin (a nerve agent)
GC	gas chromatograph, gas chromatography
GPL	general population limit

H	sulfur mustard
HAZMAT	hazardous material
HAZOP	hazardous operation
HD	sulfur mustard (distilled)
HDC	heated discharge conveyor
HRA	health risk assessment
HT	vesicant mixture: 60 percent agent H and 40 percent bis[2(2-chloro-ethylthio)ethyl] ether
HVAC	heating, ventilation, and air conditioning
JACADS	Johnston Atoll Chemical Agent Disposal System
LIC	liquid incinerator
MDB	munitions demilitarization building
MOU	memorandum of understanding
MPF	metal parts furnace
MSD	mass spectrometric detector
NARAC	National Atmospheric Release Advisory Center
NRC	National Research Council
OSHA	Occupational Safety and Health Administration
PARDOS	partial dosage
PAS	pollution abatement system
P.L.	public law
PLL	programmatic lessons learned (program and database)
PMACWA	Program Manager for Assembled Chemical Weapons Assessment
PMATA	Product Manager for Alternative Technologies and Approaches
PMCD	Program Manager for Chemical Demilitarization
PMCSD	Project Manager for Chemical Stockpile Disposal
QA	quality assurance
QC	quality control
QRA	quantitative risk assessment
RCRA	Resource Conservation and Recovery Act
RIR	recordable injury rate
SAIC	Science Applications International Corporation
SBCCOM	U.S. Army Soldier and Biological Chemical Command
SHA	systems hazard analysis
SOP	standard operating procedure
TOCDF	Tooele Chemical Agent Disposal Facility
TWA	time-weighted average
UPA	unpack area
USACAP	U.S. Army Chemical Activity Pacific
USACHPPM	U.S. Army Center for Health Promotion and Preventive Medicine
U.S.C.	United States Code
VX	a nerve agent
WCL	waste control limit
WPL	worker population limit
5X	level of decontamination (suitable for commercial release)

Executive Summary

The National Research Council was asked by the Army to form a special, ad hoc committee to investigate whether incidents involving chemical warfare materiel stored, processed, and destroyed at the two operational Army chemical demilitarization sites provide useful information for the safe operation of future sites.[1] To discharge its responsibility, the Committee on Evaluation of Chemical Events at Army Chemical Agent Disposal Facilities examined information on all forms of chemical events and incidents that occurred through the summer of 2001 at the Johnston Atoll Chemical Agent Disposal System (JACADS)[2] site in the Pacific Ocean and at the Tooele Chemical Agent Disposal Facility (TOCDF) in Utah. Information on these events was obtained from sources within the government and from a full range of public sources.

The committee concluded that safe chemical weapons disposal operations are feasible at the new facilities scheduled to begin operating at Anniston, Alabama; Umatilla, Oregon; and Pine Bluff, Arkansas, if their management is diligent in setting and enforcing rigorous operational procedures, in providing comprehensive training, in establishing a strong safety culture encompassing all plant personnel, and in absorbing programmatic lessons learned from the first two operational facilities, JACADS and TOCDF. The committee believes that many of the observations and recommendations made in this report are applicable to all demilitarization facilities, including those that may not use incineration. No evidence derived from previous chemical events causes the committee to doubt that the new incinerator technology plants or the disposal processes they will employ can be operated safely and effectively. The committee joins predecessor committees (NRC, 1994, 1997) of the National Research Council that have found that the risk to the public and to the environment of continued storage overwhelms the potential risk of processing and destruction of stockpiled chemical agent.

Recommendation 1. The destruction of aging chemical munitions should proceed as quickly as possible, consistent with operational activities designed to protect the health and safety of the workforce, the public, and the environment.

THE CHEMICAL DEMILITARIZATION CHALLENGE

How can we safely destroy the current U.S. stockpile of chemical weapons within the time constraints imposed by a dangerous and deteriorating stockpile (U.S. Army, 2001d) and mandated by law? Under congressional mandate (Public Law 99-145), the Army instituted a sustained program to destroy elements of the chemical weapons stockpile in 1985 and extended this program to destroy the entire stockpile when Congress enacted Public Law 102-484 in 1992. The stockpile then included more than 31,000 tons of nerve and blister agents deployed in several million individual munitions and containers. In 1997, the Congress reiterated this commitment by ratifying the Chemical Weapons Convention.[3]

The U.S. Army, through its Program Manager for Chemical Demilitarization (PMCD), began active destruc-

---

[1] The statement of task is included in the preface.

[2] Johnston Island, southwest of Hawaii, was the site at which the U.S. Army gathered chemical weapons withdrawn from overseas locations. JACADS, the initial stockpile facility, began destruction activities in 1990 and completed processing in November 2000. Planning for closure operations is currently under way.

[3] Formally known as the Convention on the Prohibition of the Development, Production, Stockpiling and Use of Chemical Weapons and on Their Destruction (P.L. 105-277), the CWC requires the destruction of chemical weapons in the stockpile by 2007 and any non-stockpile weapons in storage at the time of the treaty ratification (1997) within 2, 5, or 10 years of the ratification date, depending on the type of chemical weapon or on the type of chemical with which an item is filled.

tion of overseas chemical weapons stockpiles at JACADS in 1990. In 1996, PMCD commenced destruction of the continental U.S. chemical weapons stockpile at TOCDF, located at the Deseret Chemical Depot (DCD) in Tooele County, Utah. The disposal of the stockpile on Johnston Island was completed in November 2000, and by September 2001 nearly 40 percent of the chemical agent at Tooele, the site of the largest stockpile, had been destroyed. Between these two facilities, approximately 23 percent of the original chemical weapons stockpile had been disposed of by the end of the summer of 2001.

During the 10 years of JACADS operation and the first 5 plus operational years at TOCDF, a number of operational upsets or incidents occurred (U.S. Army, 2001c). Some resulted in chemical agent penetrating into normally agent-free areas where workers could be exposed. In others, improper operating procedures in agent-contaminated areas led to actual or potential worker exposure. Further, in a few of these events, very small amounts of agent were actually released outside the building into the ambient atmosphere.

JACADS and TOCDF are first- and second-generation chemical demilitarization facilities based on the disassembly of chemical munitions and destruction of both the chemical agent and the associated energetic munitions, as well as the decontamination of metal containers in a suite of specialized incinerators. In 2002 and 2003, third-generation facilities based on the same disassembly and incineration technologies are scheduled to commence operation at three of the largest remaining stockpiles at Army depots in Anniston, Alabama; Umatilla, Oregon; and Pine Bluff, Arkansas.

This report responds to congressional, Army, and public concerns by:

- Providing a context for evaluating the significance of chemical events,
- Illustrating methods for the analysis of chemical events,
- Analyzing chemical events at the two initial chemical demilitarization sites as of September 2001, and
- Providing recommendations for minimizing and managing potential future chemical events.

Dismantling and destroying chemical weapons is inherently hazardous, but the Program Manager for Chemical Demilitarization has incorporated extraordinary safety precautions into both plant design and personnel training (NRC, 1996, 1997, 1999a). The chemical demilitarization incineration plants are virtual fortresses built to withstand the consequences of accidents, and, to date, releases of chemical agent from these facilities have been rare, isolated events involving only small amounts of agent, even under upset conditions (NRC, 1996, 1997, 1999a). State-of-the-art quantitative risk assessments have determined that the major hazard to the surrounding communities arises from potential releases of agent from stockpile storage areas, not the demilitarization facilities (U.S. Army, 1996a; NRC, 1997; see also Chapter 1 and Appendix E). However, given the inherent complexity of the chemical demilitarization task at the assembled weapons stockpile sites, it is almost certain that new problems will continue to arise, particularly from aging and deteriorating weapons and the challenges of demilitarization plant closure and decommissioning. There will be future "chemical events," and serious consequences to both plant personnel and surrounding communities cannot be ruled out.

WHAT ARE CHEMICAL EVENTS?

Data and Definition

To determine the frequency and nature of chemical events at JACADS and TOCDF, the committee requested that PMCD provide information on all incidents at the two sites that the Army considered to be chemical events. PMCD provided data on 81 separate incidents (39 from June 1990 through December 2000 at JACADS and 42 from August 1996 through May 2001 at TOCDF; see Appendix B) and included independent investigation reports for the most serious events. The committee also solicited and received information on actual or suspected incidents from concerned citizens, local and state officials, an organization opposed to incineration as a disposal means, and current and former facility employees (see, for example, Appendix C). Much of this information was gathered during visits to PMCD, JACADS, TOCDF, and the recently constructed Anniston Chemical Agent Disposal Facility.

To gain a perspective on the release of chemical agent to the environment during chemical demilitarization activities, the committee obtained data from the U.S. Army Soldier and Biological Chemical Command on the rate and severity of leaks from 1990 through 2000 from the chemical weapons stockpiles stored at Johnston Island and Deseret, Utah (U.S. Army, 2001d).

The committee determined that current Army criteria for classifying events at storage and demilitarization facilities are ambiguous and allow the local depot commander latitude to define as a chemical event accidents or incidents that do not involve release of chemical agent.[4] Other incidents that clearly involved chemical agent were not defined

---

[4]For example, Army Regulation 50-6, on chemical surety, provides specific examples of chemical events which the committee judges to be so broad as to invite widely divergent interpretations by local Army depot commanders, such as example number 7: "Any malfunction or other significant activity at a chemical demilitarization plant that could reasonably be expected to cause concern within the local community or the press, or that in the judgment of the local facility or installation management or leadership could cause embarrassment to the U.S. Army" (U.S. Army, 1995).

as chemical events, because the escaped agent remained within the plant's engineering controls.

For the purposes of this report, the committee determined that a *chemical event* is any incident associated with chemical demilitarization operations that results in an actual or potential release of chemical agent.

Recommendation 2. The Army should establish a consistent set of criteria to be used by all chemical-agent-processing facilities to ensure uniformity in the classification of events, and to facilitate event analysis and comparison.

Risk Assessment

The demilitarization facilities contain relatively little chemical agent at any one time, and that agent is under stringent engineering controls in the demilitarization facility. The published quantitative risk assessment for TOCDF (U.S. Army, 1996a) makes clear that by far the greatest risk to the public arises from accidental or deliberate detonation of stored chemical munitions and the accompanying release of large amounts of chemical agent to the environment. Although after the events of September 11, 2001, the Army delayed publication of its quantitative risk assessments for the third-generation chemical demilitarization facilities, the committee has ascertained that the new risk assessments confirm the dominance of the risk of continued chemical munitions storage. The committee concluded that, in the post-September 11, 2001, world, the threat of terrorism and sabotage would likely be focused in the storage facilities, rather than the demilitarization facilities.

The committee further finds that quantitative risk assessments (QRAs) and health risk assessments (HRAs) are critical inputs to the dialogue necessary to ensure adequate public involvement in and understanding of chemical demilitarization activities. Maintaining a prudent balance between the public's right to know the risks they face and the need to protect sensitive information will be an ongoing challenge for the chemical demilitarization program. Without adequate risk information available to the public, it will be difficult to develop or maintain the level of public trust necessary for PMCD to accomplish its mission.

Recommendation 3. The Army should continue its practice of making available to the public the results of its quantitative risk assessments and health risk assessments for each chemical demilitarization site.

The committee also found that the QRAs provide a valuable framework for managing the risk from chemical events, including events arising from sabotage, terrorism, and war, by placing events in the context of their impact on safety.

Recommendation 4. The quantitative risk assessment (QRA) for each chemical demilitarization site should be iterative. Actual chemical events should be used routinely to test the completeness of the QRA, which should be routinely utilized to hypothesize the frequency and consequences of chemical events. The Program Manager for Chemical Demilitarization and the U.S. Army Soldier and Biological Chemical Command should use the QRAs to evaluate measures to control future chemical events. The Army should also consider using QRAs to examine scenarios associated with sabotage, terrorism, and war.

MONITORING CHEMICAL AGENT

The committee also reviewed the chemical agent monitoring procedures at incinerator-based demilitarization facilities. It determined that because the monitoring levels used by PMCD are very conservative and highly protective of worker and public health and safety, there are frequent false positive alarms, as well as alarms for actual events that pose no measurable threat to workers or the public. These conservative stack-monitoring thresholds ensure that no significant amounts of agent can be exhausted into the ambient air without the facility alarming and the agent incineration feed automatically terminating. In-plant air breathed by unmasked workers and the output of the scrubbing system for air exiting the chemical demilitarization plant are monitored at similarly conservative thresholds.

Recommendation 5. The Army should maintain conservative chemical demilitarization exhaust stack and in-plant airborne agent exposure thresholds. If current limits for exposure to stockpiled chemical agents are further reduced, the Army should not further reduce existing monitoring thresholds unless chemical agent monitors can be made both more sensitive and more specific so that lower thresholds can be instituted without significant increases in false positive alarm rates or unless health risk assessments demonstrate that lower thresholds are necessary to protect workers or the public.

However, the high rate of false positive alarms seems to be causing a "crying wolf" mentality whereby some operational personnel tend to discount alarms until they have been confirmed by laboratory analyses. PMCD must make it clear that properly responding to alarms is more important than production and, at the same time, show that it is trying to solve the underlying problem by actively developing better instruments. The committee notes that PMCD's operating procedures require that all alarms be treated as real until it has been demonstrated by laboratory analyses that they were not triggered by real chemical events.

Recommendation 6. To reduce the rate of false positive alarms for both airborne and condensed-materials agent contamination, the Program Manager for Chemical Demilitarization and the relevant Department of Defense research and development agencies, such as the Army Research Office, the Army Research Laboratory, the Defense Advanced Research Projects Agency, and the Defense Threat Reduction Agency, should invigorate and coordinate efforts to develop chemical agent monitors with improved sensitivity, specificity, and time response. These efforts should be coordinated with, and take advantage of, the increased level of interest in and increased resources available for developing chemical weapons detectors for homeland defense.

CHEMICAL EVENTS ANALYSES

In analyzing past chemical events, the committee found that the basic design of the incineration-based demilitarization facilities and the processes used to disassemble and destroy chemical weapons and to dispose of residue and waste streams (see Appendix A) are fundamentally sound. The committee further found that the investigation of chemical events and incidents at demilitarization facilities has been straightforward and honest. However, the committee observed that future investigations could benefit from the use of methodologies such as causal tree analysis (where events are related to the final outcome) and human factors engineering (where data on human performance are related to the causal tree). Such methodologies would result in uncovering and understanding the complete set of those factors found to have contributed to each incident.

Recommendation 7. Incident investigation teams should use modern methodologies of incident investigation routinely at all chemical demilitarization sites to help uncover a broader set of causal and contributing factors, and to enable greater understanding of the interrelationships between and among these factors. Experts in human performance modeling should be included on any incident investigation team. A standing incident review board at each site should be established to identify chemical events requiring in-depth investigation and to ensure that the lessons learned appropriately influence ongoing operations. These boards would meet regularly to review accidents and incidents, including chemical events, and would be fully informed of any findings and recommendations made by chemical event investigation teams.

In its analysis of JACADS and TOCDF chemical incidents and events, the committee observed that repeating patterns of causal factors occurred across the range of incidents, from minor to severe. In particular, deficiencies in standard operating procedures (SOPs), design failures, and understandable, although inappropriate, assumptions (mind-set) of operations personnel contributed to almost all of the incidents investigated in depth. Repeating patterns of causal factors in most incidents did not appear to have been used by management to generalize incident findings beyond the immediate context of each incident.

Recommendation 8a. The Program Manager for Chemical Demilitarization should analyze all chemical-agent-related incidents at chemical demilitarization plants for patterns of causal factors and should institute program-wide actions to address the causes found.

The programmatic lessons learned (PLL) database compiled by PMCD is a large undertaking and should help capture lessons from past chemical events and help prevent the recurrence of similar events. PMCD is to be commended for creating and maintaining the PLL database. However, information in the PLL database is relatively hard to use and is not prioritized. The data would be more useful if it were organized in a manner that included a system for prioritizing the data. The data may contain patterns that underlie several events and that could be found by "mining" the data for these connections. This information would improve the capability for broad generalization of specific information from an individual incident.

Recommendation 8b. Any improvements made in investigation procedures should become part of a systematically organized programmatic lessons learned (PLL) database that makes information easier for the non-expert to find and/or use. This can include prioritization and developing a drop-down "tree" list. Lastly, the Program Manager for Chemical Demilitarization should ensure that, at the plant level, the data are available to, known by, and useful to operations personnel. The proposed contractor for the PLL program should address these issues. For the program to be useful all stakeholders need to buy into its use and structure.

CHEMICAL EVENT IMPACTS

The committee observed that the computer models used to model accidental chemical releases in Army and local government emergency operation centers (EOCs) are representative of the state of the art as of the late 1970s. The Gaussian plume dispersion modeling techniques embedded in the D2PC computer model used to predict agent emission plume extent have more current and accurate implementations. Adoption of more modern and more accurate emission plume models seems to have been delayed by the failure to integrate better plume models into standard Chemical Stockpile Emergency Preparedness Program (CSEPP) emergency response models.

Recommendation 9a. Stockpile sites that still use the D2PC computer model should, at a minimum, upgrade their emergency response models to take advantage of the improved capabilities available in the D2-Puff model. Consideration should be given to testing and possibly optimizing the D2-Puff model at each site by performing tracer release experiments under a variety of meteorological conditions.

Recommendation 9b. The Chemical Stockpile Emergency

Preparedness Program should undertake a continuing evaluation of alternative approaches to modeling the release and impact of chemical agents.

Recommendation 9c. Accurate agent plume dispersion modeling capability should be coupled with timely communication of results and appropriate responses to the stockpile site and surrounding communities.

The committee also determined that communications during and after incidents and events have not always occurred as intended between and among the various stakeholders. The lack of an override function or a hot line dedicated to notification that an event has occurred has led to inadequate communication during chemical events. For example, the lack of notification and warning between DCD, Tooele County, and other Utah responsible agencies was caused in part by a lack of coordination between the Federal Emergency Management Agency's (FEMA's) CSEPP and the Army's Emergency Operations Center, and in part because of DCD's prevailing attitude that its emergency management responsibilities "end at the fence." This perspective, if carried to other communities where chemical demilitarization facilities are to be operated, can endanger the ability to provide an effective, coordinated emergency response to incidents. The memorandum of understanding for information exchange recently agreed to by the DCD and Tooele County (see Appendix G) could serve as a model for every community with a chemical weapons stockpile, to ensure very close oversight of the disposal plant's operations.

Recommendation 10a. Chemical demilitarization facilities should develop site-specific chemical event reporting procedures and an accompanying training program that tests and improves the implemented procedures and communication system.

Recommendation 10b. The standing incident review board recommended by the committee for each site should include a qualified member of the public who can effectively represent and communicate public interests.

Recommendation 10c. Each chemical demilitarization site should consider the establishment of a reporting and communication memorandum of understanding (MOU), of the sort developed between the Deseret Chemical Depot and Tooele County, which specifies reliable and trusted means of alerting and informing local officials about chemical events. These MOUs should be designed to permit ready evaluation and updating of the terms of the MOU to take full advantage of learning across the array of chemical demilitarization sites.

Recommendation 10d. The Army Emergency Operations Centers and the Chemical Stockpile Emergency Preparedness Program should establish a stronger capability and capacity for the coordination of training, equipment, and plans necessary to respond effectively to an emergency incident, and the commitment to do so in a coordinated and cooperative fashion. Additionally, the Army should continue its program of outreach—including listening to public concerns and responding to them, as well as engaging in more conventional public information efforts—to both the public and the relevant government oversight agencies to enhance general understanding of the chemical demilitarization program.

A major chemical event can result in several months of lost chemical munitions processing time. Multiple incident investigations and responses have led to additional delays in restarting operations when incidents have led to plant shutdown. All aspects of such investigations and resumption of operations should be accelerated consistent with safe operations.

Recommendation 11. All stakeholders and involved regulatory agencies should agree that a single team will investigate chemical events requiring outside review. This investigation team should comprise already-appointed representatives from all stakeholder groups and agencies, including members of the public who can effectively represent and communicate with local officials and the affected public. Incident findings should be documented in a single comprehensive report that incorporates the findings, proposed corrective actions, and concerns of the various oversight agencies.

ESTABLISHING A SAFETY CULTURE

The committee believes that the JACADS and TOCDF safety programs and performance have been and continue to be adequate to ensure that chemical demilitarization operations are being conducted safely. Even so, there is considerable opportunity for improvement. Many of the incidents that have occurred at JACADS and TOCDF could have been significantly mitigated—if not prevented—had a true "safety culture" been in place and functional at the time.

Recommendation 12a. Much of the needed improvement in safety at chemical weapons facilities can come from increased attention to factors that contribute to and/or cause chemical events. For example, the Program Manager for Chemical Demilitarization and chemical demilitarization facility managers should ensure that standard operating procedures are in place, up to date, and effective, performing hazard operations analyses on new process steps and design changes even when such changes are viewed as trivial and recognizing that chemical hazards are posed by things other than agent (e.g., waste).

Recommendation 12b. Management at the Tooele Chemical Agent Disposal Facility (TOCDF) and the new third-generation facilities should develop or identify and implement programs that will result in the establishment of a pervasive, functioning safety culture as well as improved safety performance. In doing so, TOCDF and the new chemical demilitarization sites should draw on experience in the chemical industry, obtained through industry associations or other appropriate venues. The Army should revise the award fee criteria to encourage each new chemical demilitarization site operator to demonstrate better safety performance than that at the older sites.

NEW FACILITY START-UP

The near-term start of operations at the three third-generation chemical demilitarization facilities presents an opportunity to get these facilities off on the right foot. Plant start-up can be a difficult learning experience for new operating crews. It is probable that conditions will arise in plant operation for which no SOP has been written. In these situations operators need an in-depth knowledge of their equipment and its limitations to handle these unusual conditions and maintain plant security. It is common practice in other industries to include "design" people in the start-up crew for new plants.

Recommendation 13. A generous allotment of time should be given to training and retraining chemical demilitarization plant operating personnel to ensure their total familiarity with the system and its engineering limitations. All plant personnel should receive some education on the total plant operation, not just the area of their own special responsibility. The extent of this overall training will be a matter of judgment for plant management, but the training should focus on how an individual's activities affect the integrated plant and its operational risk. Each facility should develop training programs using the newly designed in-plant simulators to present challenges that require knowledge-based thinking. The training programs should include a process for judging the effectiveness of the training. Including "design" experts in the start-up crew for new plants could be helpful in identifying latent failures in process and facility design.

The committee's specific findings are paired with the recommendations noted above and presented together in Chapter 6 of this report.

1

The Chemical Demilitarization Challenge

For more than 50 years the United States has maintained an extensive weapons stockpile containing chemical agents, stored primarily in military depots distributed in the continental United States. Largely manufactured 40 or more years ago, the chemical agents and associated weapons in this stockpile are now obsolete. Under a congressional mandate (Public Law 99-145), in 1985 the Army instituted a sustained program to destroy elements of the chemical weapons stockpile and extended this program to destroy the entire stockpile when Congress enacted Public Law 102-484 in 1992.

Chemical weapons stored overseas were collected at Johnston Island, southwest of Hawaii, and destroyed by the Johnston Atoll Chemical Agent Disposal System (JACADS), the first operational chemical demilitarization facility. JACADS began destruction activities in 1990 and completed processing of the 2,031 tons of chemical agent and the associated 412,732 munitions and containers in the overseas stockpile in November 2000 (U.S. Army, 2001a).

The largest continental U.S. stockpile component, which initially contained 13,616 tons of agent, is stored at the Deseret Chemical Depot (DCD) near Tooele, Utah. This component of the stockpile is being processed by the Tooele Chemical Agent Disposal Facility (TOCDF), which started operation in August 1996 and destroyed 5,320 tons of agent and processed more than 880,000 munitions and containers in its first 5 years of activity. As of September 2002, the first two chemical demilitarization facilities had destroyed over 25 percent of the original chemical agent tonnage (U.S. Army, 2002a).

The Army, through its Office of the Program Manager for Chemical Demilitarization (PMCD), now has more than 15 years of cumulative operating experience in chemical weapons demilitarization. PMCD plans to open three additional facilities in the near future to meet Chemical Weapons Convention[1] requirements for destruction of the U.S. stockpile. Despite the progress made to date, however, operations at JACADS and TOCDF have not been without incident. Several "chemical events" at the two plants have resulted in either unplanned discharge of significant amounts of agent within the facilities and/or the release of very small amounts of agent to the atmosphere above these plants.

The following sections in this chapter discuss the chemical demilitarization challenge: how to safely destroy the stockpile of chemical weapons within the available time constraints imposed by a dangerous and deteriorating stockpile. To put this challenge in context the Committee on Evaluation of Chemical Events at Army Chemical Agent Disposal Facilities describes technology for the chemical stockpile's disposal, defines and describes chemical events, discusses the significance of risk assessment to the chemical weapons disposal process, and categorizes institutional issues associated with chemical demilitarization.

STOCKPILE CONTENT, DISPOSAL DEADLINE, AND DISPOSAL TECHNOLOGY

The chemical weapons stockpile contains two types of chemical agents: the cholinesterase-inhibiting nerve agents (GB and VX), and blister agents, primarily mustard (H, HD, and HT) but also a small amount of Lewisite. Both types of chemical agents, which are liquids at room temperature and normal pressures, are frequently, but erroneously, referred

---

[1]Formally known as the Convention on the Prohibition of the Development, Production, Stockpiling and Use of Chemical Weapons and on Their Destruction (P.L. 105-277), the CWC requires the destruction of chemical weapons in the stockpile by 2007 and any non-stockpile weapons in storage at the time of the treaty ratification (1997) within 2, 5, or 10 years of the ratification date, depending on the type of chemical weapon or on the type of chemical with which an item is filled.

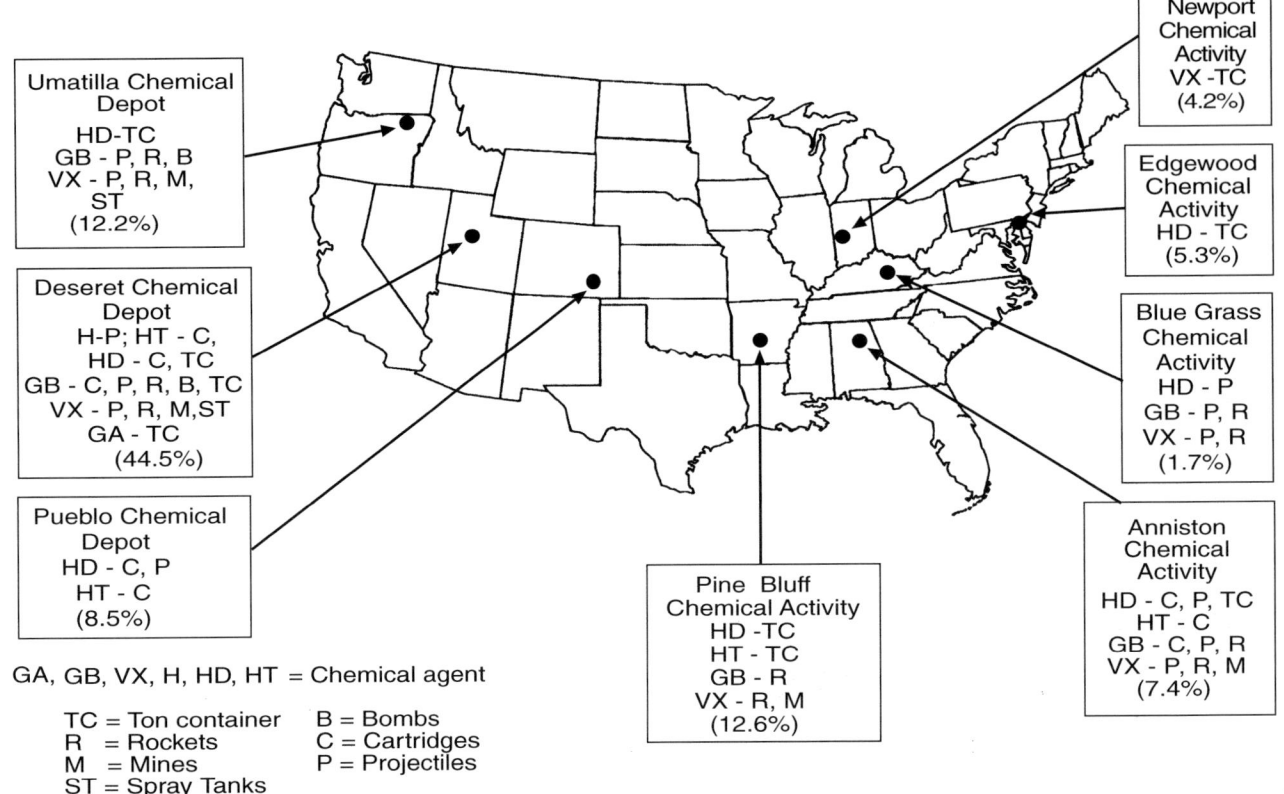

FIGURE 1-1 Location and size (percentage of original stockpile) of eight continental U.S. storage sites. SOURCE: NRC (1997).

to as gases. The stockpile contains both bulk ("ton") containers of nerve and blister agent and munitions, including rockets, mines, bombs, projectiles, and spray tanks loaded with either nerve or blister agents. Many munitions contain both chemical agent and energetic materials (propellants and/or explosives), a combination that poses particular challenges for safe and efficient destruction.

The disposal of stockpiled chemical weapons is a major undertaking. In 1990, the stockpile included 31,496 tons of chemical agents. The current stockpile is stored at eight chemical weapons depots operated by the Army in the continental United States. The location, size, and composition of the original continental U.S. stockpile is presented in Figure 1-1.

The U.S. Chemical Stockpile Disposal Program (CSDP) has evolved in parallel with international initiatives to eliminate chemical weapons. After many years of negotiation, the terms of the CWC were agreed upon in 1993 to deal with this issue. As of June 2002, the CWC had been signed by 174 countries and ratified by 145. The convention went into effect on April 29, 1997, after ratification by 65 countries. The CWC requires that signatories, which include the United States, destroy their chemical weapons stockpiles within 10 years of its initiation, making April 29, 2007, the deadline for destruction of the U.S. stockpile. A provision in the treaty allows a 5-year extension of the deadline under some circumstances. As of early October 2001, PMCD released new schedule estimates indicating that chemical demilitarization activities at the three disposal facilities scheduled to commence operation in the near future may not be completed until 2008 at Pine Bluff, Arkansas, and until 2009 at Anniston, Alabama, and Umatilla, Oregon (U.S. Army, 2001b).

The disposal technology selected by the Army for storage sites that contain a full range of chemical agents and munitions types is a multifurnace incineration process (NRC, 1999a). In this "baseline" technology approach munitions and containers are drained of agent, which is burned in dual liquid incinerators (LICs). Robotic machinery disassembles munitions containing energetic charges and the separated energetic materials are burned in a rotary kiln-based deactivation furnace system (DFS). Sheared bulk containers and metal munitions parts are fed though a large heated metal parts furnace (MPF) designed to burn off any residual agent or energetic material, decontaminating metal components to the point that they can be recycled as normal scrap metal. The LIC, DFS, and MPF furnaces are all equipped with extensive pollution abatement systems (PASs) designed to substantially eliminate gaseous and particulate exhaust material of potential concern and exhaust remaining gases through a common stack.

The first-generation incineration system was deployed at JACADS. The second-generation system, deployed at

TOCDF, is described in detail in the National Research Council report *Tooele Chemical Agent Disposal Facility—Update on National Research Council Recommendations* (NRC, 1999a)[2]; this detailed description is reprinted in Appendix A. Third-generation incineration systems are currently close to operational status at the Pine Bluff, Anniston, and Umatilla sites. The new facilities will use basically the same process as that used at TOCDF and JACADS. Weapons will be taken apart in the same way, and there will be the same three lines of incineration: a rotary furnace for destroying propellant and explosive materials (see Appendix A, Figure A-6), a furnace with a moving conveyor primarily for decontaminating metal parts (see Figure A-7), and a furnace for burning liquid agent (see Figure A-8). Improvements to the new facilities have been made compared with TOCDF and JACADS, however; these are noted in Chapter 5.

In addition, the Army has selected liquid-phase hydrolysis processes, supplemented by various secondary hydrolysate treatment and/or disposal processes, to destroy the chemical agents contained only in bulk "ton" containers at Newport Chemical Depot in Indiana and at the Edgewood Chemical Activity site on Aberdeen Proving Ground in Maryland. Significant problems in introducing these new technologies to dispose of even the simplest case of "bulk only" chemical agent have been recognized (NRC, 2000a). A disposal technology has not yet been selected for the relatively small stockpile of nerve and mustard munitions at the Blue Grass Chemical Depot in Kentucky. At the Pueblo Chemical Depot in Colorado the Department of Defense has decided to use neutralization, followed by bio-treatment of the secondary waste to dispose of the mustard munitions stored there.

CHEMICAL EVENTS

During the 10 years of JACADS operation to destroy the chemical weapons stockpile at Johnston Island and the first 5 years of operation of TOCDF, a range of operating incidents occurred that were designated as chemical events by the local Army depot commanders. Army Regulation 50-6 on Chemical Surety (U.S. Army, 1995) defines chemical events very broadly: "The term chemical event encompasses all chemical accidents, incidents and politically/public sensitive occurrences." The regulation goes on to give specific examples, such as:

1. Confirmed releases of agent from munitions outside a closed containment system, such as a filtered bunker, storage igloo, or overpack container.

2. Discovery of an actual or suspected chemical agent container or munition in a place where it is not supposed to be that may require emergency transportation or disposal.
3. Confirmed detection of agent above the threshold concentration for any period outside the primary engineering control.
4. Actual exposure of personnel to agent above the allowed limits specified in various Army regulations.
5. Loss of chemical agent.
6. Any terrorist or criminal act directed toward a chemical agent storage, laboratory, or chemical demilitarization facility or any deliberate release of chemical agent.
7. Any malfunction or other significant activity at a chemical demilitarization plant that could reasonably be expected to cause concern within the local community or the press, or that in the judgment of the local facility or installation management or leadership could cause embarrassment to the U.S. Army.

At the eight continental U.S. storage sites, the Army's local depot commander has the responsibility to decide whether an upset or incident within the storage yard or at the associated chemical demilitarization facility is a chemical event. Examples 1 through 6 above seem to imply that, in most cases, chemical events are those in which chemical agent ends up where it should not be, i.e., in the ambient atmosphere or under the control of an unauthorized individual. However, no such requirement is inherent in example 7. The wide latitude in judgment about what might "cause concern within the local community or the press, or . . . could cause embarrassment to the Army" that is delegated to the depot commander suggests that some incidents defined as chemical events by one commander may not be considered chemical events by another.

Whatever the local Army commander deems to be a chemical event is subject to strict reporting procedures detailed in Army Regulation 50-6 (U.S. Army, 1995). Both telephone reports (within 3 hours) and initial written reports following specified formats (within 24 hours) must be made to both the Army Operations Center and Headquarters–Department of the Army. These reports are usually shared with local authorities and serve as the basis for press releases issued by the local depot and/or PMCD, but there appear to be no general guidelines for the form and the timing of such notification. A tabulation of chemical events is also provided in the Army's annual reports to Congress summarizing chemical agent storage and chemical demilitarization activities (see, for example, U.S. Army, 2000a).

Chemical Events Associated with Disposal

The Program Manager for Chemical Demilitarization (PMCD) is required to prepare and provide reports of chemi-

---

[2]This update report also details TOCDF technology and management issues identified by an NRC oversight committee during that facility's first 3 years of operation.

cal events for incidents so designated that occur within its facilities, as specified in Army Regulation 50-6.[3]

PMCD provided to the committee a chronicle of 81 incidents that occurred over the 10 years of JACADS operation and the initial 5 years of TOCDF operation (U.S. Army, 2001c). As shown in the Army's document (see Appendix B), a significant number reported under example 7 listed above did not involve chemical agent. Of those events involving chemical agent, only a few resulted in release of agent outside of engineering control and into the atmosphere. The total mass of chemical agent released to the environment in these incidents was almost certainly less than a gram (U.S. Army, 2001d), which is equivalent to no more than a few drops. The committee's analysis of PMCD-reported chemical events at the two chemical demilitarization facilities is presented in Chapter 2.

In addition to a list of PMCD-reported chemical events, the committee also received lists of possible chemical-agent-related incidents from local officials and concerned citizens groups (see Appendixes C and D). These are also addressed in Chapter 2.

Chemical Events Occurring During Storage

A listing and analysis of the chemical incidents involving leaking containers and munitions at the Johnston Island storage site from 1990 until the end of disposal operations in 2000 and at the Tooele site (now the DCD) from 1990 through 2000 were provided to the committee by the Army's Soldier Biological and Chemical Command's (SBCCOM's) Stockpile Management Team (U.S. Army, 2001e). As a result of its continuous stockpile inspection program SBCCOM has records of the frequency of chemical agent leaks occurring in stockpiled munitions and containers. Most of the incidents listed by SBCCOM involved a single leaking munition or container, although incidents involving more than one leaking munition discovered in a storage igloo were not uncommon. One incident involved 20 leaking munitions treated over the course of a month. The most serious incidents, including all those known to have discharged a significant amount of agent outside of engineering control, were designated as chemical events and reported as required by Army Regulation 50-6.

According to the SBCCOM statistics on stockpile leakage at Johnston Island, 13 incidents involving leaking munitions were reported from 1990 through 2000. Ten of these occurred from 1990 through 1992, with only 3 occurring in 1993 or later. Some of the later falloff is likely due to reduction of the stockpile as chemical demilitarization proceeded, with the most problematic munitions scheduled for the earliest destruction, according to the overall risk mitigation strategy outlined below in this chapter. Inspection and remediation of corroded or leaking munitions prior to their shipment to Johnston Island probably also contributed to the fact that so few munitions and containers leaked in the Johnston Island storage site.

The statistics for stockpile leakage at the Tooele site (DCD) for the same period from 1990 to 2000 differ considerably from those for JACADS. SBCCOM tabulated 31 incidents for 1990, 34 for 1991, 40 for 1992, 37 for 1993, 38 for 1994, 33 for 1995, 26 for 1996, 14 for 1997, 14 for 1998, 10 for 1999, and 11 for 2000, for an 11-year total of 288 (U.S. Army, 2001e). Again the record of events suggests that as stockpile destruction proceeds, with the most problematic weapons and containers scheduled for early destruction, the stockpile leakage statistics improve. From 1990 through 1995, 72 percent (159) of the 213 tabulated stockpile leakage incidents involved GB, which was the agent type destroyed at TOCDF through March 2002.

Some of the 288 storage chemical agent incidents at DCD from 1990 through 2000 involved only relatively low storage igloo vapor readings with most, possibly all, of the released agent captured in the carbon air filters present in igloos storing the highest-risk munitions (generally those containing GB) (U.S. Army, 2001e). In more that 100 different incidents, however, storage personnel who entered an igloo reported observing liquid agent leaks with volumes estimated to range between 1 teaspoon and 2 quarts, with one leak from a GB ton container totaling 10 gallons. In dozens of additional incidents smaller amounts of leaked liquid were observed. Since most of these incidents involved high-vapor-pressure GB, many resulted in release of small amounts of agent to the atmosphere, if only when the igloo door was opened to allow entry. Many of these leaks were initially detected by monitoring igloo air drawn through sampling ports. Mobile powered air filtration units were often used to minimize agent migration out of the igloo.

In addition, in 11 incidents at DCD from 1990 through 2000, ton containers of mustard, stored outside, were found to be leaking agent directly into the environment (U.S. Army, 2001e). When these incidents occurred, storage site personnel attempted to quantify the amount of agent lost by estimating the volume of contaminated gravel or soil underneath the leaking container, or, in the case of large leaks, measuring the agent remaining in the container. For most of the outdoor incidents documented at DCD, relatively small amounts of agent (a few drops to a few cups) were estimated by depot workers to have been lost. However, in the most serious event, a leak of distilled mustard estimated at 78 gallons (~375 kg) was discovered on September 9, 1993. The volume of agent released in this incident alone swamped the total mass of known emission of agent from chemical demilitarization facilities by at least a factor of several hundred thousand.

Large outdoor releases from storage facilities are an ongoing concern. In fact, while the committee was gather-

---

[3]Detailed guidance on the preparation and distribution of these reports and on associated record keeping is presented in a periodically updated document designated PMCD Regulation 385-3, "Accident and Chemical Event Notification, Investigation, Reporting and Record Keeping" (U.S. Army, 1999a).

ing data for this report during a visit to TOCDF on July 26, 2001, a leaking plug in a ton container of HD in the DCD storage site produced a vapor plume large enough to force workers at adjacent TOCDF to don respirators. That leak, according to chemical event reports and related memoranda supplied by SBCCOM, was determined to be about 9 pounds (~4 kg). This incident delayed the committee's access to TOCDF for several hours.

While chemical demilitarization operations at JACADS and TOCDF have released small amounts of chemical agent into the environment, these releases are negligible compared with releases to the environment from associated chemical weapons storage sites. The rate of agent leaks and releases does decrease significantly as the stockpile is processed.

TOOLS FOR ASSESSING HAZARDS IN THE OPERATION OF CHEMICAL STOCKPILE DISPOSAL FACILITIES

The Army has developed a suite of risk assessment and risk management tools to permit analysis of potential risks in terms of the scenarios that can contribute to risk, the likelihood of those scenarios, and the consequences associated with them. Those consequences are as follows (NRC, 1997, p. 16):

> For humans (both workers and the public) there are three potential measures of risk either from the stockpile or from stockpile destruction: acute lethality; acute and latent noncancerous health effects; and latent cancer. The potential adverse consequences for the environment are the contamination of land and/or water and adverse effects on native or endangered species.

These tools can be used to evaluate the risk associated with specific chemical events. Real-world events can also then be used as a check on the analyses, enabling revision of risk analyses to include new classes of events when surprises occur.

The variety of analysis tools is useful because of the differing needs of various program elements. To understand how they are related, the committee first groups these tools into two large classes: prospective (or predictive) tools and retrospective (or documentation) tools.[4]

Prospective Risk Analysis Tools

Health Risk Assessment

A health risk assessment (HRA) is a compliance-oriented analysis that examines the risk to a set of stylized receptors (e.g., the subsistence fisherman) associated with routine releases (intended to be conservative upper bounds based on tests and performance of other units) and mild upset conditions (assumed to lead to release of a multiple of the conservative routine release for a specific fraction of the year). Accidents, specific systems failures, and specific human actions are not considered. The HRA is an upper-limit risk estimate for routine operations. Because it does not provide a realistic estimate (accounting for uncertainty), does not consider accidents, and does not address worker risk, it is not helpful in evaluating chemical events, other than providing a baseline, of sorts, against which the consequences of chemical events can be evaluated. For an example HRA see the analysis of TOCDF sponsored by the Utah Division of Solid and Hazardous Waste (Utah DEQ, 1996).

Systems Hazards Analysis

A systems hazards analysis (SHA) is a systematic and comprehensive search for and evaluation of all significant failure modes of facility systems components that can be identified by an experienced team. The hazards assessment often includes failure modes and effects analysis, fault tree analysis, event tree analysis, and hazards and operability studies. Generally, the SHA does not include external factors (e.g., natural disasters) or an integrated assessment of systems interactions. However, the tools of SHA are valuable for examining the causes and the effects of chemical events. They provide the basis for the integrated analysis known as quantitative risk assessment. For an example SHA see the TOCDF Functional Analysis Workbook (U.S. Army, 1993-1995).

Quantitative Risk Assessment

A quantitative risk assessment (QRA) is an integrated, quantitative analysis (including uncertainty) of accident scenarios, their likelihood, and possible consequences. Current QRAs examine human actions as well as systems failures, external events as well as internal failures, and worker risk as well as public risk. A salient feature of a QRA is that it is integrated, in that it:

- considers the interactions of systems and their effects on each component, considers common causes of failures, and considers all forms of system dependencies
- considers the integrated impact of multiple system and human failures on the potential for releases
- considers the impacts of weather and emergency protection on public consequences

Thus, the QRA provides an effective tool for evaluating the significance of chemical events. In fact, scenarios leading to chemical events and the frequency and consequences of these events are exactly what a QRA describes and calculates. Real-world events provide a check on the analysis. If

---

[4]The committee uses the Army's names and acronyms for these methods. Use of these names is not consistent with language in other environments.

potentially risk-significant events occur that were not previously modeled, the QRA can and should be updated to account for that event and any similar events that could occur. For an example QRA see the Army's TOCDF QRA (U.S. Army, 1996a). The committee notes that the TOCDF QRA was the first PMCD QRA and does not include all the features in the current analyses being finalized for the facilities at Anniston, Umatilla, and Pine Bluff. At the time of this writing the TOCDF QRA is the only one that has been published. Similar QRAs are being completed at the remaining sites. It is possible that portions of these may be unavailable publicly because of security concerns.

Key elements of the Army's approach to quantitative risk assessment are summarized in Appendix E for the interested reader. More details are available in the NRC Stockpile Committee's risk report (NRC, 1997).

Retrospective Analysis Tools

Monitoring Systems

Monitoring systems detect releases of hazardous chemicals, providing warning of hazardous conditions and a record of their occurrence and extent. They can also measure the burden of chemicals on the human body. They are not predictive, but instead provide real-time observations. For a description of monitoring schemes, see Box 1-1 and the NRC Stockpile Committee's report *Occupational Health and Workplace Monitoring at Chemical Agent Disposal Facilities* (NRC, 2001a).

Chemical Event Investigations

When chemical events occur, investigations identify what actually happened and when, the reasons, and the consequences; they usually suggest corrective actions for the future. An investigation (separate from possible corrective action) is most effective when it focuses on what actually happened from the viewpoint of those involved (i.e., why the actions of people involved made sense to them at the time, what they could see and what they knew, how they viewed their alternatives) (Weick and Sutcliffe, 2001). Too often these investigations are biased by hindsight and focus on what the operators might have done rather than why they did what they did. An effective investigation identifies the organizational and management issues that made the actions seem reasonable to those involved, and it can provide a basis for real improvement. Chemical event investigation and analysis are the subjects of Chapter 2 of this report.

Putting It All Together

From this brief introduction, it is clear that the QRA, chemical agent monitoring, and event investigation are the key tools for addressing the safety issues associated with chemical events. In its published QRA the Army performed a detailed assessment of the risk of public fatalities and cancers associated both with the stockpile storage sites and chemical weapons processing activities at Tooele (U.S. Army, 1996a), and it has performed ongoing risk assessments for the planned third-generation incineration system chemical demilitarization facilities and associated stockpile storage areas at Pine Bluff, Anniston, and Umatilla. In its QRA for TOCDF, the Army's analyses indicated that, over the facility's projected operating schedule, the risk associated with accidental releases of agent due to disruption of the stockpile, most likely due to earthquake or leaks from ton containers of GB, greatly outweighed the risk of release of agent due to chemical demilitarization activities (U.S. Army, 1996a). This risk assessment does not examine potential terrorist activities, threats that are addressed by other federal agencies in addition to the Army.

The Army's risk assessment for TOCDF and its associated storage facility was reviewed by the NRC and found to be sound (NRC, 1997). Even in the event of an earthquake or plane crash that damages the disposal plant, the risk of public fatalities due to release of agent from the disposal facility is calculated to be about 5 percent of the expected risk of fatality due to releases of agent from the storage yard (U.S. Army, 1996a; NRC, 1997). A more detailed discussion of the TOCDF QRA and of advances incorporated in subsequent QRAs is presented in Appendix E.

Until the last few days of the disposal schedule, the amount of agent in the storage yard greatly exceeds the amount in the chemical demilitarization plant; as the stockpile is depleted, the risk posed by the storage facility drops proportionally. A key risk management strategy adopted by the Army is to order the stockpile destruction so that the most volatile, highly toxic agent and associated munitions are processed first (those containing the nerve agent GB), while less volatile and/or deadly agents are processed later.

Finally, it is important to note that the original TOCDF QRA focused on public risk, and little effort was devoted to examining worker risk. One consequence of this limitation in scope was that very little modeling of human performance was done in the TOCDF QRA. As attention in the program shifted to include worker risk, more significant modeling of human action has been performed. None of these improved analyses have yet been published. A variety of human reliability analysis methods have been used (Gertman and Blackman, 1994). For ongoing work, new approaches that account for details of context and human cognitive function are being adapted (Hollnagel, 1998; USNRC, 2000). With more careful and complete analysis, new scenarios especially important to worker risk are being developed. These methods, integrated into the risk assessments, can be used to quantify the impact of human actions on situations posing risk. Human performance not only is a significant component in risk assessment, but also, as the committee learned in its study, is directly involved in most chemical events. In

> **BOX 1-1   Details on Airborne Chemical Agent Monitoring Methods and Standards at Chemical Demilitarization Facilities**
>
> Two systems are currently used to monitor concentrations of airborne chemical agents at chemical demilitarization facilities. One system, the automatic continuous air monitoring system (ACAMS), is designed for "near-real-time" monitoring (currently ~3- to 8-minute cycle time, dependent on agent, for a single instrument). The ACAMS consists of an air sampling system connected to a gas chromatograph (GC) equipped with a flame photometric detector (FPD).
>
> Specific columns and detector filters are used for each agent. The nerve agents, GB and VX, are detected by phosphorus oxide chemiluminescence, due to their P content, excited in the FPD detector, while mustard is detected by sulfur dimer chemiluminescence, from its sulfur content. Since VX has high molecular weight (298 amu), it is catalytically cleaved at the entrance to the GC column to shorten its detection time. This detection scheme relies on the characteristic GC column transit time of the agent or agent fragment (in the case of VX) plus the P or S spectrally specific flame chemiluminescence detection signal to identify the agents. The method is quite sensitive; ACAMS are often run at threshold detection volume-mixing ratios of a part per trillion (pptv) or lower. However, at these low threshold levels false positive alarms often occur because other chemical species can "interfere" by producing chemiluminescent signals that overlap the time gate and spectral band pass associated with the agents. For time-critical applications, like exhaust stack monitoring, the GC cycle time can be mitigated by time phasing two or more ACAMS sampling the same gas stream.
>
> ACAMS alarms must be verified to ensure that they are not a false positive due to an "interferent" species or instrument malfunction. This verification is done using a depot area air monitoring system (DAAMS) deployed near an ACAMS. DAAMS is a passive system that draws an air stream through a sorbent tube. The tubes are collected and replaced periodically if there are no ACAMS alarms or shortly after an alarm occurs. They are transported to a laboratory and thermally desorbed onto a sample tube and analyzed on a laboratory scale GC/FPD system. Without confirmation by the more sensitive and specific laboratory GC/FPD system, the ACAMS alarm is not confirmed. If the laboratory GC/FPD system does not show a chromatogram consistent with agent, a second DAAMS sample may be run on a laboratory GC equipped with a mass spectrometric detector (MSD). The GC/MSD analysis is designed to identify interferent compounds that may have caused a false positive ACAMS alarm.
>
> The Army currently mandates very conservative alarm thresholds for it chemical demilitarization facilities (U.S. Army, 1997a; NRC, 2001a). Current exhaust stack alarms are set at 0.2 of the allowable stack concentration (ASC), for GB the ASC is just three times the time weighted average (TWA), which serves as the worker population limit (WPL) for the demilitarization workforce. The TWA is the level of agent an unmasked person can breathe for an 8-hour shift without harm. Thus, GB, the most volatile agent and therefore the greatest airborne threat to surrounding communities, is monitored at stack concentrations equivalent to 0.6 of the level currently deemed safe for a worker to breathe for a full shift without protection.
>
> Stack exhaust monitoring levels for the less volatile and less threatening HD and VX are monitored at levels a factor of 2 and 6 above their TWA (WPL) levels, respectively. The 0.2 ASC stack level for GB is a factor of 10,000 below the "immediately dangerous to life and health" (IDLH) level for this agent. In-plant air levels breathed by unmasked workers and the output of the scrubbing system for air exiting the demilitarization plant are monitored at similarly conservative levels (generally 0.2 TWA). Since any agent in either the stack exhaust gas or scrubbed plant air will be greatly diluted before reaching the facility's fence line, air flowing into the surrounding communities will be well below the "general population limit" (GPL) defined as the level believed to pose no threat to the public. The GPL for the three stockpile agents is set at 30 to 33 times lower than their TWA (U.S. Army, 1997a; NRC, 2001a). The current ASC, TWA, and GPL levels for GB are 3×10^{-4}, 1×10^{-4}, and 3×10^{-6} mg/m^3. For VX these values are 3×10^{-4}, 1×10^{-5}, and 3×10^{-6} mg/m^3, while for HD they are 3×10^{-2}, 3×10^{-3}, and 1×10^{-4} mg/m^3 (NRC, 2001a; U.S. Army, 1997a).

Chapter 2, the committee examines the more significant of the chemical events at JACADS and TOCDF to determine their characteristics with respect to facility performance and human performance. How these events are related to safety performance is not a simple question. In his widely referenced book (Reason, 1997), in a chapter devoted to the relationship between frequent, low-consequence events and the risk of high-consequence events, James Reason concludes that:

> If both individual and organizational accidents have their roots in common systemic processes, then it could be argued that . . . personal injury statistics are indicative of a system's vulnerability (or resistance) to organizational accidents. The number of personal injuries sustained in a given time period must surely be diagnostic of the "health" of the system as a whole. Unfortunately, this is not so. The relationship is an asymmetrical one. An unusually high [personal injury rate] is almost certainly the consequence of a "sick" system that could indeed be imminently liable to an organizational accident. But the reverse is not necessarily true. A low . . . rate (of the order of 2-5 per million man hours)—which is the case in many well-run hazardous technologies—reveals very little about the likelihood of an organizational accident.

The problem of human-caused events, how to control them, and how to discern the difference between high- and low-risk events continues to be studied in many industries (Reason, 1997; Hollnagel, 1998; IOM, 2000).

Monitoring Methods

The occurrence and the extent of a release of chemical agent are tracked through PMCD's workplace chemical agent monitoring system as described in NRC (2001a). Monitoring for airborne agent is a major activity at each chemical agent disposal facility. Box 1-1 provides details on monitoring.

Sensitivity requirements for the near-real-time automatic continuous air monitoring system (ACAMS) for airborne agent are demanding. This is because the allowable stack concentration and time-weighted-average levels used for exhaust stack and in-plant action levels are quite low and because the ACAMS alarms are currently set at 0.2 of the relevant action level.

This demand for sensitivity results in relatively frequent false positive alarms, particularly for the ACAMS monitoring the individual incinerator exhaust flows and the common exhaust stack (NRC, 1999b). Previous NRC reports have noted that the frequency of false positive ACAMS alarms disrupts plant operations, particularly when stack alarms trigger an automatic shutdown of agent feed to the liquid incinerator, and can lead to an unsafe "crying wolf" mind-set that tends to discount ACAMS alarms (NRC, 1999b, NRC, 2001a). In fact, one of these studies found evidence that the May 8-9, 2000, agent stack release at TOCDF was exacerbated by an expectation that what proved to be real exhaust system ACAMS alarms were instead just false positives (NRC, 2001a). While previous NRC reports have urged PMCD to improve both the reliability and time response of its airborne agent monitoring systems (see NRC, 1999b for a summary), progress in this area has been modest.

Another weakness of the airborne monitoring system is the lack of real-time (<10 seconds) agent detection. The NRC has previously recommended that the Army develop a real-time system that uses a measurement technology independent of the gas chromatography with flame photometric detector methods used by the ACAMS and the depot area air monitoring system (DAAMS) (NRC, 1994). To date, the Army's attempts to develop and demonstrate such a real-time system have not been successful (NRC, 1999b, 2001a). New interest in chemical agent detection as a key component of antiterrorism activities has spurred government and commercial activities focused on developing better sensors for airborne agent (IOM, 1999). The NRC has also urged the Army to continue to monitor technological advances in trace gas detection and to consider implementing any that are appropriate for monitoring agent in chemical munitions disposal facilities (NRC, 1999b). Renewed interest in chemical agent detection and monitoring methods spurred by homeland defense concerns may lead to better and more robust technology. The committee urges PMCD to vigorously seek out and exploit any suitable developments arising from these activities.

Previous NRC reports have also noted the lack of robust techniques for rapidly measuring agent and agent breakdown products present in liquid waste streams and associated with solid materials (NRC, 2000a; NRC, 2000b; NRC 2001a). These reports recommend vigorous efforts to develop better methods to measure agent contamination in these media.

Event Analysis and Significance

The committee notes the importance of chemical event analysis that focuses on the viewpoint of the operators during the sequence of events. Understanding why their actions seemed appropriate to them, *at the time*, is the key to effecting real improvement in performance. Gaining an understanding of the factors within their work environment—training, equipment, and operational indications, as well as goals and rewards—which led them to conclude that their actions were appropriate is an essential element of developing a real safety culture at the facility.

An associated effort is to ensure that the QRA includes the class of events that actually have occurred. Mapping real event scenarios onto scenarios modeled in the QRA allows one to see a particular action integrated into the larger system for each chemical event and thus determine its effect on safety.

CHEMICAL DEMILITARIZATION INSTITUTIONAL ISSUES

Trust and Institutional Arrangements

The chemical demilitarization program necessarily depends on a combination of trust and institutional arrangements to accomplish the destruction of the chemical stockpile. Because extremely hazardous materials and complex technologies are involved, those seeking destruction of chemical agent and munitions must rely on agencies and firms expert in these processes to carry out the chemical demilitarization program. In essence, legislation and regulatory agency rule making establish institutional and contractual arrangements for the chemical demilitarization program, stipulating what is to be accomplished and (in some cases) how it is to be done. As in any contract, the "principal" relies on an "*agent*"[5] to accomplish a task or service, and provides

---

[5]It is unfortunate that use of the term "agents" to indicate those that carry out tasks for "principals" might in this report be a source of confusion in the context of the chemical demilitarization program (where "agent" usually refers to chemical agent). Where *agent* is used in this report in the institutional sense, it is italicized to reduce the potential for confusion.

the means to ensure that the task is accomplished according to the principal's needs (Wood, 1992; Scholz and Wei, 1986). The U.S. Congress and the public it represents rely on agencies of the U.S. Army, state and federal regulators, private contractors, and a host of other entities to carry out the chemical demilitarization program. At specific chemical demilitarization sites, the local public and the officials who represent them similarly depend on these *agents* to carry out the task safely and effectively.

When principals delegate complex tasks, they create a relationship in which the *agents* on whom they rely are more knowledgeable about the task than are the principals. *Agents* design, test, construct, operate, and modify the chemical demilitarization facilities and have intimate knowledge of these steps, while principals often rely on *agents* for such knowledge. This kind of information asymmetry may place the principal at a disadvantage in overseeing the safety and effectiveness of the program, and necessitates monitoring and control mechanisms that are specified in the relevant laws and contracts. Monitoring mechanisms include permitting and reporting requirements, inspections, investigations, and rules governing whistle-blowers, while control mechanisms include arrays of incentives (such as contract fee structures) and sanctions (civil and criminal punishments, fee deductions, and so on). A trade-off implicit in this relationship is that of the principal's control over *agents* versus the scope of the *agent's* discretion, some of which is typically necessary for complex and demanding tasks that require the *agent* to push the boundaries of known processes and technologies. Greater trust reduces the need to rely on formal monitoring and control, and conversely loss of trust increases the need for monitoring and controls.

The Institutional Setting of Chemical Demilitarization

The U.S. government's approach to chemical demilitarization involves a complex amalgam of institutional stakeholders. The Army's SBCCOM is the operator of the eight remaining stockpile storage facilities. PMCD is responsible for the construction, operation, and subsequent closure of JACADS, TOCDF, the three new incinerator system facilities at Anniston, Umatilla, and Pine Bluff, and the two-bulk only, hydrolysis-based facilities under construction at Newport and Edgewood (Aberdeen). By law, evaluation of alternative (non-incineration) technologies that may be used to dispose of the stockpiles located at Pueblo, Colorado, and Blue Grass, Kentucky, has been delegated to an independent Program Manager for Assembled Chemical Weapons Assessment (PMACWA) within the Army.

Protection of the public from harm due to accidental releases of agent near storage depots and associated chemical demilitarization facilities is the responsibility of the Chemical Stockpile Emergency Preparedness Program (CSEPP), which is funded by the Army but administered by the Federal Emergency Management Agency (FEMA).

Thus, at any of the six continental sites where a chemical demilitarization facility is either operating or under construction, a concerned citizen needs to receive a consistent and accurate message from a range of state and federal entities including PMCD, SBCCOM, FEMA, and CSEPP. In the past a consistent message from these Army or Army-related entities has been hard to achieve (NRC, 1999a).

In addition, chemical demilitarization facilities must obtain environmental permits from state environmental agencies in order to commence operations and must seek permit amendments and renewals from these same agencies in order to sustain operations. Permit conditions may vary widely from state to state even though the state environmental agencies operate largely under authority delegated from and overseen by the federal Environmental Protection Agency. State-to-state discrepancies in chemical demilitarization facility operating permits or amendments to existing operating permits may raise public concerns. Hearings related to environmental permit applications and amendments give citizens an important opportunity for input into the operation of chemical demilitarization plants.

The PMCD receives guidance from the Army Medical Services (U.S. Army Center for Health Promotion and Preventive Medicine), working in conjunction with the Department of Health and Human Services' Centers for Disease Control and Prevention, about the levels of exposure to chemical agent that are considered safe for both workers and the public (U.S. Army, 1990, 1991). Recent reevaluations have led to proposals for significantly lower recommended standards related to exposure to chemical agents. This possibility has raised citizen concern about the safety of Army stockpile storage and chemical demilitarization operations designed to meet the current exposure limits.

Chemical events have raised questions about the safety of the stockpile storage and the demilitarization process. Understanding whether an event results from flaws in design, fundamental problems with technologies, organizational failures, or personnel lapses is essential for determination of appropriate responses. Because the answers to these questions materially affect the circumstances of the *agent*, concerns about whether *agents* are sufficiently forthcoming and responsive are inevitable. The U.S. Congress has responded to such concerns with diligent oversight (including the request for this report), requirements that whistle-blowers be protected from retaliation, and requests for formal annual reports from the Army on the progress of chemical demilitarization and the occurrence of any chemical events associated with either chemical demilitarization or storage facilities.

REPORT ROADMAP

The chemical events that have occurred at JACADS and TOCDF are characterized and a selected subset analyzed in Chapter 2. Chapter 3 discusses protocols and processes for reporting chemical events, outlines how selected events were reported at both facilities, and discusses how these events affected plant operator interactions with other stakeholders, including environmental regulators, elected state and local officials, and the public. Chapter 4 discusses the implications of lessons learned from past chemical events and their impact on continuing operations at TOCDF and future operations at Anniston, Umatilla, and Pine Bluff. Prudent preparations to minimize the occurrence and impact of future chemical events at incineration system chemical demilitarization facilities are discussed in Chapter 5. Chapter 6 contains focused findings and recommendations drawn from material presented in the first five chapters.

2

Causal Factors in Events at Chemical Demilitarization Facilities

Early in its deliberations, the committee recognized that different stakeholders have different perceptions of what constitutes a chemical event. It further became apparent that the sheer number of incidents recorded for JACADS and TOCDF made a detailed review of each event beyond the committee's resources and time. To focus its efforts, the committee identified from the full list of incidents compiled by a variety of groups (see Appendixes B, C, and D) a comparatively small number of serious events that could be evaluated in some detail. The committee's goal was to select representative occurrences so that this report's findings and recommendations would be generally applicable.

This chapter describes the committee's process for defining a chemical event, its rationale for selecting which of the large number of chemical events it would analyze in depth, and what its analysis of operational events inside each facility determined.

DEFINITIONS

One of the first issues addressed by the committee was what constitutes a chemical event. The Army's definition of chemical events encompasses all chemical accidents, incidents, and politically and publicly sensitive occurrences (U.S. Army, 1995), whether or not chemical agent was actually present. The committee determined that the seven examples provided in Army Regulation 50-6 (U.S. Army, 1995; see Chapter 1) were too broad for the tasks assigned to it. Consequently, it elected to establish its own criteria to determine which of the reported incidents qualified as chemical events.[1] The following definition was developed by the committee and used for the selection process:

Chemical event: Any incident associated with chemical demilitarization operations that resulted in an actual or potential release of chemical agent.

As used in this report, the term "release" refers to agent detected and confirmed in an area where agent is not normally present or expected to be present. Further, as described in this report, an "environmental release" refers to agent detected and confirmed in the environment outside the chemical demilitarization facility. Additionally, the committee had an interest in whether there was "worker exposure" involved in the chemical event.

SOURCES OF INPUT AND SELECTION OF EVENTS FOR IN-DEPTH ANALYSIS

Any analysis of events must recognize a continuum of potential events, ranging from expected and safe variations of processes to serious events that harm people or damage equipment. If too narrow a set of events is chosen for analysis (for example, only those with severe consequences), patterns of contributing factors may be difficult to identify. Conversely, too broad a set of incidents includes much "normal" variation that merely confirms that process controls are functioning as planned. The amount of effort devoted to the investigation of events tends to be a function of the severity of the outcomes, with the result that much more detailed data are available on the (rare) major events.

The committee received written or verbal communication from stakeholders and/or their representatives describing a large number of potential chemical events.

- The Army Program Manager for Chemical Demilitarization (PMCD) provided a written list of 81 events (Appendix B) that occurred after operations began at Johnston Atoll Chemical Agent Disposal System (JACADS) and at Tooele Chemical Agent

---

[1] The committee's purpose in reclassifying chemical events was solely to assist in selecting the events that it would review, and not to "second-guess" the Army's classification system.

Disposal Facility (TOCDF), as well as detailed investigation reports on several of the incidents.
- The Calhoun County (Alabama) Commissioners provided a letter detailing concerns and questions for the committee and including a list of six chemical events and a number of areas of concern (Appendix D).
- The committee met with Congressman Bob Riley (R-Ala.) at his request, and with representatives from Calhoun and Talladega counties, plus concerned citizens and governmental officials from Alabama, at a Capitol Hill meeting arranged by Congressman Riley to provide the committee with a local perspective.
- The Chemical Weapons Working Group (CWWG) provided the committee with a list of 118 items (Appendix C). Several committee members discussed some CWWG concerns with Craig Williams, the executive director of the CWWG, at the Capitol Hill meeting.
- A verbal presentation was made and submitted in writing by Gary Harris, a former employee and whistle-blower at the Chemical Agent Munitions Disposal System (CAMDS) facility and at TOCDF, at the committee's meeting of October 18, 2001.
- A verbal presentation was made by Suzanne Winters, chair, Utah Citizens Advisory Commission (formerly science advisor to the governor of Utah), at the committee's meeting of October 18, 2001.
- A set of 69 Notices of Violation at TOCDF issued by the State of Utah's Department of Environmental Quality, Division of Solid and Hazardous Waste, on February 13, 2001, was reviewed.
- A subgroup of the committee visited Anniston, Alabama, and received comments from local officials and citizens.

Of these submissions, the three formal lists of events supplied to the committee (by PMCD, the Calhoun County Commissioners, and the CWWG) had some events in common that are discussed further below. The written submission by Gary Harris focused principally on his experiences at the CAMDS facility, which was not part of this study.

The PMCD Incident List

The PMCD provided to the committee a list of 81 incidents, 42 at TOCDF and 39 at JACADS (U.S. Army, 2001c; see Appendix B). The Army had classified 24 (17 at TOCDF and 7 at JACADS) of these as "chemical events." Of the 81 incidents, some were significant enough to warrant investigation by agencies external to the incineration facility. The committee obtained investigation reports for 14 of the incidents and supplemented the information in them by interviewing managerial, operating, and laboratory personnel during site visits to JACADS and TOCDF. The committee also obtained data from process logs and other operational documents to assist with detailed analysis of specific incidents. Using its agreed-to definition of a chemical event and drawing on the extensive reports, the committee reevaluated this extensive material and designated 40 events (19 at TOCDF and 21 at JACADS) as chemical events.

To focus its analysis, the committee decided to examine events with the following characteristics: (1) sufficient investigation had already been done to provide a basis for analysis and (2) the event could have had potentially serious outcomes, was complex in nature, was well documented, and provided a rich source of potential causal factors. With this as a rationale, the committee examined five dissimilar incidents in significant detail (Table 2-1).

The committee then analyzed two relatively recent events, both of which resulted in the release of agent into the environment and triggered detailed investigations (Table 2-2; see Boxes 2-1 and 2-2 for details on the two events).

The Calhoun County Commissioners' List

The Calhoun County (Alabama) Commissioners submitted a letter (see Appendix D) that listed six areas of concern about operations at TOCDF. Those concerns included six chemical events the commissioners wished the committee to evaluate. They also requested that the committee evaluate events described or concerns raised by groups of concerned citizens. The only citizen group that provided such a listing was the CWWG.

Five of the incidents identified by the commissioners were included in the PMCD incident list (Appendix B) and were reviewed either in the committee's overall examination or in its detailed analyses; the remaining incident could not be confirmed as having happened. Many of the other concerns expressed by the commissioners were deemed to be outside the scope of the committee's statement of task, although some, such as the operation of the chemical agent monitoring systems and the potential impact of changes in demilitarization technology and/or operational procedures, are examined in this report. To ensure that a full range of possible incidents was considered, members of the committee met in person with the Calhoun County Commissioners at their offices on December 3, 2001, to discuss their concerns within the constraints imposed by NRC committee guidelines.

The Chemical Weapons Working Group Incident List

The Chemical Weapons Working Group provided a list of 118 items to the committee (see Appendix C), 55 of which were notations of operational shutdowns and unconfirmed automatic continuous air monitoring system (ACAMS) alarms, for example: "Site masking alarm and/or stack alarm. Potential case of chemical warfare agent release or release of other related toxic chemicals (unidentified to date)." It is probable that most, if not all, of the "site masking" alarms

TABLE 2-1 Events on the PMCD List That Were Examined by the Committee

Date	Demilitarization Site and Army Classification	Process Component	Incident / Event Description by PMCD
21-Jan-92	JACADS (Unusual Occurrence)[a]	Deactivation furnace system (DFS)	Processing VX-filled M55 rockets when a detonation occurred within the DFS, causing the kiln to stop rotating.
2-Jan-93	JACADS (Unusual Occurrence)	Explosive containment room (ECR)-A	During M60 105-mm projectile processing within the ECR a fire occurred along the miscellaneous parts conveyor. Fire was contained within the ECR. Changes made to the equipment and increased frequency of ECR cleanup of residual explosives.
17-Mar-93	JACADS (Chemical Event)	Munitions demilitarization building (MDB)	Ratheon Engineering and Constructors worker potentially exposed to mustard agent (HD). Worker developed blister(s) on leg after handling HD-contaminated waste materials.
23-Mar-94	JACADS (Chemical Event)	Common stack	Liquid incinerator (LIC) was being ramped down (controlled cooling operation) for slag removal. Minute amount of GB released via common stack. Technical investigation completed and operation procedures changed.
19-Nov-94	JACADS (Unusual Occurrence)	ECR	Detonation of rocket on fuze shear caused agent migration to observation corridor. All agent vapor contained under engineering controls and exhausted through the MDB charcoal filter units.

[a]The committee's definition of a chemical event requires that the event result in actual or potential release of agent *in an area where agent is not normally present or expected to be present*. The committee categorized the January 1992, January 1993, and November 1994 incidents as unusual occurrences because no agent was released or migrated to areas where it was not supposed to be, and further, the potential of this happening was considered slight. Conversely, the March 1993 and March 1994 incidents were categorized as chemical events because both resulted in the release of agent into the environment.

SOURCE: Excerpted from U.S. Army (2001c); see Appendix B.

TABLE 2-2 Events on the PMCD List That Were Chosen by the Committee for Detailed Analysis

Date	Demilitarization Site and Army Classification	Process Component	Incident / Event Description by PMCD
8-May-00	TOCDF (Chemical Event)	Deactivation furnace system (DFS)	During processing of GB rockets the DFS interlock shut off all burners due to pollution abatement system air flow meter failure. ACAMS alarmed in the furnace stack during re-light of the furnace. No agent or munitions were being processed at time of the alarms. The perimeter monitors' readings were all negative for agent. Investigation teams from CDC (Centers for Disease Control and Prevention), Department of Army Safety, and Utah DSHW (Division of Solid and Hazardous Waste) conducted the investigation of stack release. Technical investigation completed with recommended procedural and design changes.
3-Dec-00	JACADS (Chemical Event)	DFS waste bin	Chemical agent (VX) was detected and confirmed in the ash from the heated discharge bin at the DFS. The agent was detected during routine monthly sampling for metals as required by the RCRA (Resource Conservation and Recovery Act) permit. The bin was isolated and placed under engineering control, and subsequently the bin was fully enclosed under engineering control.

SOURCE: Excerpted from U.S. Army (2001c); see Appendix B.

> **BOX 2-1 December 3-5, 2000, Johnston Atoll Chemical Agent Disposal System (JACADS) Event**
>
> The destruction of the last agent-containing munitions on Johnston Island, M23 VX land mines, was completed on November 29, 2000. This marked the end of the operational phase of JACADS and the beginning of the closure phase. One of the first steps of closure was to process bulk solid waste (items such as spill pillows, rags contaminated with explosive or agent, metal hardware, rubber hoses, etc.) from the explosive containment room (ECR) through the deactivation furnace system (DFS). The material was processed using the standard 5X procedure (1000°F for 15 minutes) and the ash and unburned material produced placed in disposal bins. A bin was sampled monthly for agent analysis.
>
> Between 7:47 PM on Dec 2, 2000 and 12:56 AM on Dec 3, 2000, three spill-pillows (each containing approximately 20 pounds of liquid waste) were processed. How much of that was chemical agent VX is unknown. The spill-pillows contained talcum powder and an amorphous silicate absorbent. The 5X treated remains of the pillows, cardboard mines, fuses, and kicker chutes passed through the DFS and the non-combustible ash exited the heated discharge conveyor (HDC) to bin 135. At 8:06 AM on Dec 3, 2000 bin 135 was placed in the staging area (outside primary engineering control) with the lid open to cool.
>
> At 10:30 AM on Dec 3, 2000, a routine sample of the solid waste from bin 135 was taken for waste control limit (WCL) analysis and the bin lid closed. The analysis (12:30 AM Dec 4, 2000) indicated a suspected interference. An extraction analysis on the same sample confirmed the presence of VX at 3000 WCL at 1:56 AM Dec 5, 2000. A second sample was taken at 3:00 AM Dec 5, 2000 and analysis indicated 5045 WCL. At 4:30 AM Dec 5, 2000, bin 135 and two others were moved to the unpacking area for further monitoring.
>
> At 10:10 AM Dec 5, 2000, an automatic continuous air monitoring system (ACAMS) reading of 1476 time weighted average (TWA) was measured in air drawn from the bottom of bin 135. After another positive ACAMS reading, the site alarm sounded at 10:20 AM and all personnel were masked and sent to checkpoint "Charlie" for possible evacuation. Depot area air monitoring system (DAAMS) confirmation of VX in bin 135 was obtained at 3:00 PM Dec 5, 2000. The hazardous materials (HAZMAT) team began a series of checks of all other bins at 12:13 PM Dec 5, 2000 and found all readings less than TWA. The DFS kiln was restarted at 9:19 PM Dec 5, 2000 to maintain a negative pressure in the HDC waste bin enclosure. An all-clear was sounded at 9:39 PM. No agent was measured at the perimeter DAAMS tubes throughout the incident.
>
> The Chemical Event Report was submitted within 3 hours of the event and the JACADS field office and U.S. Army Chemical Activity Pacific made notifications to their respective field offices. The Program Manager for Chemical Demilitarization (PMCD) made telephone notifications to the Assistant Secretary of the Army for Installations, Logistics, and the Environment, the Department of the Army Safety Office and the Department of Health and Human Services, however no notification was given to Region IX, Environmental Protection Agency. PMCD initiated an investigation to protect evidence and gather information and assembled an investigation team on Johnston Island on Dec 13, 2000.
>
> The conclusions of the investigation team as summarized in the report were: "The process of sending VX contaminated liquid and saturated spill pillows to the DFS in excess of the decontamination capability of the furnace system appears to be the major cause of the chemical event. There are no other scenarios consistent with the physical evidence observed in bin 135 that could have resulted in the agent levels that were recorded during this chemical event. A faster response from the lab and a procedure that includes an action level for the exceedance of waste control limits would have reduced the amount of time bin 135 was outside of engineering controls. A detailed review of standard operating procedures for bulk solid waste fed to the DFS should be conducted. In addition, a narrower definition of what constitutes bulk solid waste should be developed."
>
> SOURCE: Reprinted from U.S. Army (2001f).

noted were false positive ACAMS alarms, which are discussed in some detail in Chapter 1. Thirty items were simple statements of fact that bore no relationship to the committee's task, for example: "August 1, 1997—Former Chief Safety Officer, Steve Jones is ruled for in his Dept. of Labor Wrongful Termination Action. Judge awards Jones his job back and $500,000 or no rehiring and $1 million. Judge calls EG&G managers liars." Four items appeared to be related to stockpile storage, and not to chemical demilitarization operations. Seventeen of the items on the CWWG list were identifiable as being related to incidents or events included on the PMCD list and were considered by the committee.

For most of the items on the CWWG list, no specific documentation or details were included beyond one to a few sentences. The committee concluded that the majority of the items were not germane to its statement of task. Those that were relevant were typical of the ones from the PMCD list that the committee studied intensively. In conclusion, the committee determined that evaluation of additional items on the CWWG list would not materially influence the findings and recommendations of this report.

BOX 2-2 May 8-9, 2000, Tooele Chemical Agent Disposal Facility (TOCDF) Event

During processing of rockets containing the chemical agent GB, at approximately 4:20 PM on May 8, 2000, a jam occurred in the lower feed gate of the deactivation furnace system (DFS) feed chute from the explosives containment room (ECR). Operators sprayed water into the chute in an attempt to clear the feed gate jam. The last of the material in the furnace had cleared the DFS and the heated discharge conveyor (HDC) by 5:30 PM. At approximately 6:10 PM the pressure was lowered in accordance with non-normal operating plans. An alarm indicating high air flow rates through the DFS and pollution abatement system went off at 8:20 PM and by 8:42 PM pressure fluctuations were affecting the operation of the DFS induced draft fans.

Meanwhile, at approximately 8:30 PM, personnel entered the area to inspect the feed chute and found enough debris to fill a coffee can. The decision was made to wash down the chute. With several openings and closing of the feed gates and spraying with water, the pressure controlling equipment was unable to stabilize the pressure in the kiln. The DFS operator took manual control in attempting to stabilize the pressure. The wash down of the chutes was completed by about 9:30 PM. The maintenance personnel then changed the strainers in ECR - B and placed approximately one pound of agent contaminated waste on the upper feed gate (this was the source of the agent that eventually was monitored in the stack, but the operators were unaware of its presence). The DFS operators continued to have difficulty stabilizing the furnace system. About 10:00 PM the DFS burners were automatically shut down and operators locked out by a malfunction signal sent by the DFS exhaust flow meter.

While seeking approval to by-pass the lock out of the burners and restart the afterburner, the common stack automatic continuous air monitoring system (ACAMS) alarmed at 11:26 PM. The site was immediately masked. A depot area air monitoring system (DAAMS) tube was taken for analysis at 11:38 PM and another put in its place. ACAMS readings as high as 3.63 allowable stack concentration (ASC) were obtained. The furnace was "bottled up" (dampers closed to slow airflow) at 11:44 PM. By 12:18 AM on May 9, 2000 the ACAMS had cleared and the order to unmask given.

Restarting of the DFS afterburner was attempted again at 12:23 AM, but the furnace went to a negative pressure and fluctuated once again. Another burner lockout occurred this time because the clean liquor pump was not running. At 12:28 AM, the DFS duct ACAMS alarmed and the site was masked again and the furnace was "bottled up" at 12:32 AM. The alarm cleared and the site was unmasked at 1:07 AM. DAAMS tubes from the perimeter were collected around 6:55 AM and subsequent analysis showed no detectable agent. The analysis of the stack DAAMS tubes indicated a stack release of 18-36 mg.

The TOCDF control room notified Deseret Chemical Depot (DCD) emergency operations center (EOC) at 11:30 PM on May 8, 2000 following the stack ACAMS alarm and updated the report at 11:42 PM with the highest readings and the fact that the duct ACAMS had also alarmed. They further notified the DCD EOC at 12:25 AM on May 9, 2000 that all ACAMS had cleared and that DAAMS analysis was pending. At 12:32 AM the DCD EOC was informed that the stack ACAMS were back in alarm and at 1:17 AM that DAAMS tubes from the first set of alarms confirmed the presence of agent GB.

At approximately 3:00 AM on May 9, 2000 notification was made by the DCD EOC to the Utah Department of Environmental Quality (DEQ) and at approximately 3:34 AM to the Tooele County dispatcher. The event was classified as a Limited Area Event (not likely to leave the site). No action was taken by the state or county until normal business hours on May 9.

Investigations were conducted by the TOCDF contractor EG&G, the Army Safety Office, the Centers for Disease Control and Prevention (CDC), and the Utah DEQ. Suspension of agent burning was initiated and stayed in effect until corrective actions recommended by the reports were made and approved by the Utah DEQ. The CDC report concluded that there was neither an impact to the health of TOCDF workers nor the general public. Subsequent computer modeling indicated that no harm to humans would occur beyond 8 ft. past the top of the 200-ft. common stack.

Resumption of operations in the two liquid incinerators and the metal parts furnaces (none of which were involved in the event) followed approval on July 28, 2000. Approval to resume operations in the DFS was given September 29, 2000.

SOURCE: Compiled from Utah DEQ (2000a), U.S. Army (2000a,b), and CDC (2000).

Notice of Violation Reports

The Notice of Violation reports issued by the Utah Department of Environment Quality (DEQ) for TOCDF contained a total of 69 items. These often differ in nature from the events listed by PMCD and others, in that they were mainly failures to observe and follow prescribed procedures, and, in general, did not lead to chemical events. Table 2-3 shows the frequency of occurrence of each type of violation reported by DEQ.

Although many of these violations were classified by the committee as minor, they are important as indicators of

TABLE 2-3 Committee's Classification of 69 Items Cited in Notice of Violation Reports

Violation Type	Number
Operational error (wrong feed, missed analysis, use of faulty equipment)	20
Failure to test/inspect on schedule	13
Failure to follow plans/procedures/specifications	11
Failure to keep correct records	7
Improper storage	5
Storage time limits exceeded	5
Incorrect labeling of waste	2
Failure to notify of changes	2
Other	4

systemic operating problems. Record-keeping errors or instances of exceeding time limits for testing or inspection, which tend to occur in all complex processes, may be indicative of insufficient resources devoted to the tasks to be performed, or lack of priority setting to prevent such "minor" infractions. The committee considered each of these as it developed its findings and recommendations.

ANALYSIS OF SELECTED CHEMICAL EVENTS

The committee's analysis was conducted on several levels. First, members investigated the causal factors for each of the seven events listed in Tables 2-1 and 2-2. They then developed a notional causal tree for each of the two events in Table 2-2 that were analyzed in depth. For illustrative purposes, a causal tree developed by the committee for the December 3-5, 2000, incident at JACADS appears at Appendix F. The tree is a standard tool in reliability analysis and is particularly useful in human reliability analysis where operator actions contribute either positively or negatively to an incident. Lastly, the committee provides a series of general and specific observations about the events.

Causal Factors

The committee's analysis of the seven chemical events listed in Tables 2-1 and 2-2 showed that there were multiple causal factors for all of the selected events. (Note: the committee could determine causal factors only for incidents for which sufficient investigation data were available.) Rather than being specified for each incident, the causal factors identified by the committee are grouped into the following generic categories:

- *Standard operating procedure (SOP) deficiencies,* including nonexistent SOP(s), inadequate SOP(s), and SOP(s) being circumvented or ignored as a routine operating practice. Such deficiencies contributed to 6 of the 7 incidents subjected to in-depth review (Table 2-4) and were noted as being involved in at least 14 of the incidents that received less thorough review by the committee. Note also that 11 of the 69 items in the Notice of Violation reports (see Table 2-3) involved similar failures to follow procedures. Several incidents involved multiple SOP deficiencies, and in one, the March 17, 1993, incident in which a worker was exposed to HD, at least six SOP deficiencies were noted, including:
  —No procedures for loading/handling bags.
  —Placing HD sludge in plastic bags.
  —Tagging bags improperly.
  —No pre-entry hazards briefing.
  —Improper carrying of bags.
  —Failure to wear proper personal protective equipment.

Following existing SOPs could have prevented several of the incidents that occurred at both TOCDF and JACADS. However, the non-compliance with SOPs was not a question of operators being contrary. Most operators were in fact trying to smooth or simplify the process by using non-approved methods, and had presumably been reinforced in this approach by past experiences. SOPs are not always perfect, for example, in that they apply to conditions not quite met at the particular time they are required. If the safe alternative is to stop work whenever an SOP is not exactly appropriate, that may not always be apparent to the operator.

- *Failures of communication,* including failure to communicate essential information, failure to heed communicated information, and inadequate communication systems, contributed to four of the incidents reviewed in-depth by the committee, and to at least five others. The March 17, 1993, and May 8, 2000, incidents could have been prevented had communications failures not occurred. In the March 17, 1993, incident, the supervisor of the work group noted that a bag containing HD waste was leaking and communicated this information to the individual handling the bag. The warning was not heeded, and subsequently the contents leaked onto the individual who was carrying the bag. In the May 8, 2000, incident, the control room supervisor was not informed that the agent strainer was to be changed during a demilitarization protective ensemble entry to clear the lower feed gate, or that the agent-contaminated strainer was being placed on the gate. During the course of this event, at many points the operator performed actions that were later seen to have been unfortunate. This suggests that the design of the system displays was not adequate to obtain an integrated overview of what was happening. This fact was recognized after the incident and a new single-screen display was developed to assist operators. However,

TABLE 2-4 Frequency of Causal Factors in the Seven Incidents Analyzed by the Committee

Date	Causal Factors						
	SOP Deficiencies	Communication Failure	Unexplained Human Error	Equipment Malfunction	Design Deficiency	Improper Technique	Mind-set
21-Jan-92					1		1
2-Jan-93	1					1	1
17-Mar-93	6	1	1	1	1	2	1
23-Mar-94	1				4		2
19-Nov-94	1	1	1	1	3		1
8-May-00	2	2		1	1	1	1
3-Dec-00	3	1			1		2
TOTAL	14	5	2	3	11	4	9

during the committee's visit, the operator and supervisor took about 10 minutes to find this screen, suggesting that it is not often used. Also, when the screen was located, it was found to be an all-text display, rather than an analog or pictorial representation. All-text displays are good for obtaining detailed information but poor for obtaining an integrated view of changing situations or conditions. The implication is that the fix was not a great improvement over the existing system.

- *Unexplained human error* is a category that describes human actions that were wrong for no reason recorded in the investigation reports or for which there is no apparent explanation. One example is the operator who assembled a piece of equipment incorrectly. The committee suspected that a more complete investigation would reveal causes for such errors.
- *Equipment malfunction* refers to the failure of equipment to function as designed but does not include design deficiencies. Contributing to three of the seven incidents subjected to in-depth review, and to at least nine other incidents, these failures ranged from simple tearing of waste bags to breakdowns of critical instrumentation such as flow meters and sensors. The committee noted that in virtually every incident involving equipment malfunction, there was a precursor, for example, installation of a flow sensor on the wrong side of a water flow control solenoid (design deficiency).
- *Design deficiency* applies to equipment or facilities found to perform operating functions inadequately as a result of their poor design. In several incidents examined by the committee, entrainment of agent into nonagent areas by personnel leaving a demilitarization protective ensemble entry could have been avoided if a timed interlock had been designed into transitional airlocks to ensure sufficient purging of airlock. Design deficiencies were found to have contributed to six of the seven incidents reviewed by the committee and to at least five others. Although a higher frequency of design deficiencies might be expected in the early phases of an operation, this does not appear to have been the case for either TOCDF or JACADS—at least based on the information that was available to the committee. The committee notes, however, that one of the chemical events it examined was directly attributable to failure to capture and implement at TOCDF design changes made at JACADS.

- *Improper technique* refers to a manner of performing tasks that causes either a hazard or a malfunction. An example is using equipment for purposes other that those dictated by design, as occurred in the May 8, 2000, incident at TOCDF in which the water spray nozzles designed for cooling the deactivation furnace system (DFS) lower feed gate were used to clean the gate when jams occurred. Since the nozzles were operated at low pressure, operators used significant quantities of water in attempts to clean or clear the feed gate and the water vaporized, causing fluctuations in pressure and in the flow rate in the DFS. While these factors were not frequent, they contributed to several incidents.
- *Mind-set* refers to the mental attitude people have about the process of disposal and the state of the system during processing. In the incidents studied, people behaved at times as if they assumed that an ACAMS alarm was false, that contaminated waste was less hazardous than raw agent, or that parts coming through a furnace were automatically 5X material.[2] During its review of incidents, the committee

---

[2]5X refers to a level of decontamination at which solids may be released for general use or sold (e.g., as scrap metal) to the general public in accordance with applicable federal, state, and local regulations. There is a misconception that 5X means simply that the solid has been placed in a temperature zone of 1000°F or higher for 15 minutes. To achieve a 5X level of decontamination a solid must be heated to 1000°F and maintained at that temperature for 15 minutes.

invariably found itself engaged in discussions of the mind-set(s) prevalent at the time of the incident(s). Mind-set was involved in every incident the committee reviewed in depth, and it contributed significantly to several others. Perhaps the most troubling was what the committee referred to during its deliberations as the "false positive mind-set." False positive ACAMS alarms have been frequent at both JACADS and TOCDF and have caused people at both sites to assume that any alarm without a readily apparent cause is false—an assumption that has, in turn, fostered other failures and delays in addressing and responding to events.

Table 2-4 summarizes the results of the committee's analyses, indicating the frequency with which the causal factors outlined above contributed to the severe incidents closely examined by the committee.

Causal Tree Analysis of Two Events

For the two events it examined that were sufficiently documented to allow a detailed analysis, the committee charted activities in the sequence of events leading to each incident, either as a time line or as a causal tree (see Appendix F). A standard tool in reliability analysis, the causal tree or event tree is particularly useful in analyzing incidents to which operator actions contribute either positively or negatively. Figure F-1 in Appendix F shows the causal tree for the December 3-5, 2000, event at JACADS. The committee recognizes that such trees are designed at the discretion of the analyst and should not be construed as reflecting scientific certainty. Figure F-1, as well as a similar analysis by the committee of the May 8-9, 2000, event at TOCDF, suggests that the incidents examined by the committee grew from normal activities into potentially dangerous events.

The activities charted can be categorized as ranging from normal operations through system response. In addition, some can extend back in time before the occurrence of the incident, e.g., latent failures.

- *Normal tasks*—that the system was attempting to accomplish before the adverse event occurred. Examples are maintenance and operations.
- *Latent failures*—conditions present in the system for some time before the incident, but evident only when triggered by unusual states or events. Examples include equipment design deficiencies, unexpected configurations of munitions, or routine ignoring of standard operating procedures.
- *Active failures*—events *before* which there were no adverse consequences and *after* which there were. Active failures are usually the result of personnel decisions or actions. These same actions may have resulted in safe outcomes on previous occasions, but in the incidents examined by the committee, such actions combined with latent failures to cause some adverse consequences. Examples of active failures include use of the wrong procedure, incorrect performance of an appropriate procedure, or failure to correctly and rapidly diagnose a problem.
- *Immediate outcome*—the adverse state the system reached immediately after the active failure. Examples are release of agent, plant damage, or personal injury. Reporting and investigation flow charts supplied by the Army indicate that the severity of outcome often determines the incident's prominence for managers, the workforce, or the community, which in turn drives subsequent responses. Incidents with more salient outcomes naturally receive more scrutiny, which may bias the data set used for analysis.
- *System responses*—actions taken to correct the effects and anticipate the aftereffects of an adverse outcome. Following each event, there is a system response that also needs to be analyzed. How did the system for incident response function? How did the management act to improve safety? Was an exposed worker properly treated? Were communities notified appropriately? How did the plant return to a normal state? How rapidly did it return? Finally, how was the system changed in light of the incident? This stage of analysis is considered in Chapter 4.

General Observations

Based on its review, the committee believes that the chemical events and other serious incidents examined at JACADS and TOCDF have been honestly investigated and reported. Even so, the investigation reports that were available to the committee did not always reflect the complete set of factors that caused or contributed to the cause of events. Likewise, the investigation team(s) may not have used the most appropriate methodologies for collecting, analyzing, and reporting the events. In particular, the committee saw little evidence of the use of formal methods, such as event tree analysis, and little involvement of human factors engineering even though most of the incidents reviewed by the committee had a component of human behavior as a causal factor (see Table 2-4). The committee found inconsistencies in the form and format of investigation reports within and between chemical demilitarization sites.

Finally, the committee noted that complete documentation supporting incident investigations was not always retained with the reports or in a report file. For example, a videotape relevant to the December 3, 2000, incident at JACADS could not be located for the committee to view.

During its in-depth review, the committee observed differences in the types and completeness of entries made in JACADS and TOCDF operating logs (deactivation furnace system, demilitarization protective ensemble, control room, and so on). The variations were largely attributable to individuals who made the entries, which suggests that some training relative to the nature, content, and detail of entries into operating logs would be appropriate. Error-correction deficiencies were also noted in the operating logs.

Specific Observations

In conducting its detailed examination, the committee observed patterns of causal factors or categories of activities, such as latent and active failures, that appeared to recur over significant time periods. Deficiencies in standard operating procedures, which can be readily identified and corrected and should decline with time and operating experience, were the most notable. Based on the information available to the committee, it appears, however, that the frequency of SOP deficiencies in the incidents examined did not decline with time. This might suggest that any lessons learned from past experience are being interpreted too narrowly (Chapter 4) or that the need for improvement in this area is not being recognized. As noted earlier, following an SOP may not appear to be the correct choice to an operator. This is particularly true when the operator has a limited perspective on the task and so does not understand the reasons why a procedure that looks unnecessarily complex is indeed appropriate. This circumstance argues again for operator knowledge in addition to rule following.

As in any complex system, there are likely still undetected design deficiencies at TOCDF, and, most certainly, systemization at new chemical demilitarization facilities will uncover other design deficiencies. Active communication between and among chemical demilitarization facilities via the programmatic lessons learned (PLL) program (Chapter 4) is key to ensuring that design deficiencies are detected and corrected.

Equipment failure may be random, but it is certainly preventable. Excellent maintenance, equipment monitoring, and preventive maintenance practices can dramatically reduce equipment malfunctions at a lower overall cost than that incurred in an unanticipated shutdown. Many industries have found that investment in these practices can provide reductions in overall costs.

Equally, human errors are preventable, even if they appear to be random. Better knowledge of human functioning in complex situations (human factors engineering) shows how equipment design, workforce knowledge, and management environment can contribute to human error, or to its reduction (Reason, 1997). Industry experience has shown that a well-trained and vigilant workforce, and vigorous and effective management and supervision, committed to creating an environment in which safety is always first, will help to minimize human errors and any ensuing events that might be caused or initiated by them. Similarly, the human component of failures in communication and improper tech-

---

BOX 2-3 An Example of Negative Effects of Mind-set

The committee highlights a sentence in an investigation report that begins the section titled "Air Monitoring of 5X Material": "The waste located in BIN 135 was designated 5X by the process. Therefore, there was no requirement to monitor for an airborne agent hazard" (U.S. Army, 2001f).

Although it agrees that the process had been demonstrated to be capable of producing 5X [decontaminated] material, the committee asserts that the waste bin enclosure should have been actively monitored to ensure that 5X destruction was being achieved on a continuous basis. To the committee, this case is not different from that of the liquid incinerator, where "6 nines" destruction efficiency has been demonstrated but does not obviate the need for monitoring to ensure that the operating requirements are achieved. It was known that certain materials could pass through the deactivation furnace system without complete combustion (e.g., rolled-up coveralls), and thus, the operating assumption regarding 5X decontamination was known to be erroneous in some cases. This assumption also led to employees being sent on two occasions to deal with the waste bin with an inappropriate level of personal protective equipment, and the "false positive" mind-set led to delays in reporting the results of monitoring.

---

niques can be greatly reduced, if not eliminated, through the development of a strong safety culture in the chemical demilitarization work environment.

The "crying wolf" phenomenon of a decreased willingness to respond after repeated false alarms is an expected, and sensible, human behavior, but one that must be discouraged in chemical demilitarization operations by appropriate training and a recognized reward structure.

The committee also discussed "waste mind-set"—the attitude or belief among employees and management at both JACADS and TOCDF that waste processing and/or handling is less hazardous than agent processing. This mind-set has led to notable deficiencies in SOPs for waste handling and contributed significantly to several incidents. Even though mind-set cannot be considered to be the root cause of any of the incidents reviewed by the committee, it is a prevalent factor (see Table 2-4) and a significant issue, as the December 3, 2000, deactivation furnace system waste bin incident at JACADS illustrates (U.S. Army, 2001f) (see Box 2-3).

The most difficult challenge facing those operating future demilitarization facilities will be overcoming, or preventing the development of, mind-sets that lead to an adverse chemical event or contribute to the severity, magnitude, and consequences of such an event. This challenge is also important to bear in mind as sites transition from agent disposal operations to decommissioning and closure.

3

Responses to Chemical Events at Baseline Chemical Demilitarization Facilities

Concentrating on the procedures for reporting and disclosing events and the legal processes involved, in this chapter the committee reviews onsite investigations and reports triggered by the chemical events discussed in Chapter 2 to determine if general conclusions can be drawn about whether those responses can assist in the tasks of determining the causes of events and preventing their recurrence. The committee concentrates on the two events involving release of chemical agent to the environment analyzed in detail in Chapter 2—the December 3-5, 2000, incident at Johnston Atoll Chemical Agent Disposal System (JACADS) and the May 8-9, 2000, incident at Tooele Chemical Disposal Facility (TOCDF) (see Boxes 2-1 and 2-2)—both of which triggered detailed investigations.

The committee also examines how emergency response professionals estimate the potential population exposure from a chemical event, reviews emergency response activities and public responses, and discusses how the events are communicated to local news media and interested citizens groups. These communications have important implications, since they affect how political leaders, regulators, and the general public view the chemical demilitarization program.

FORMAL EVENT REPORTING PROTOCOLS

Formal protocols for reporting a chemical event establish a communication network designed to alert the chemical demilitarization facility staff and plant workforce and the surrounding community to any imminent danger and to mobilize emergency assistance in case of a major event. Additionally, there are a variety of reporting requirements to the Army, the Environmental Protection Agency (EPA), and state and local emergency operations centers, as well as reporting protocols within the facility operating contractor's organization and the Program Manager for Chemical Demilitarization (PMCD) organization.

Generally, the first indication of a problem is an automatic continuous air monitoring system (ACAMS) alarm, but because many interfering chemicals also cause an alarm, declaration as a chemical event requires laboratory confirmation by the more accurate depot area air monitoring system (DAAMS) analysis (which can take from 20 minutes to more than an hour).[1] If an ACAMS alarm is confirmed within the chemical demilitarization facility, the installation commander must be notified. Army Regulation 50-6 requires installation commanders to notify the Army Operations Center by telephone within 3 hours of the time a chemical event is confirmed and in writing within 24 hours. A confirmed event must further be reported to EPA within 24 hours (U.S. Army, 1996b). PMCD has tailored the Army's regulations to support its mission and requires notification within 1 hour of confirmed events.

The Army Materiel Command (AMC) has set additional guidelines for reporting incidents, including those that (1) have a potential for negative reactions from local officials or the media, (2) involve workers reporting possible exposure to agent, and (3) involve detection of agent outside primary engineering controls but within secondary engineering controls. The state and local protocols for any given plant are determined on a case-by-case basis in accordance with state and local regulations and laws.

Located on an isolated island, JACADS had only EPA Region IX to report to at the local level. Contingency procedures for dealing with agent outside engineering controls were approved in the early days of the project and included a flow chart and call-down lists. The contingency plans involved notification of on-site U.S. Army Chemical Activity

---

[1] Incidents triggering ACAMS alarms that are not verified by DAAMS analyses are considered to be Resource Conservation and Recovery Act (RCRA) events that require reporting within 15 days.

Pacific (USACAP) soldiers, the Johnston Island Fire Department, the Johnston Island airport, and resident personnel. Priority was placed on basic notification of fires, explosions, agent releases, and serious bodily injury. There was a call-down list, and a written log was kept. Military officials in Washington, D.C., were notified within 24 hours.

Two Army reporting chains run in parallel. The *green suit* chain culminates at the Chief of Staff of the Army and the *civilian* chain with the Secretary of the Army. For chemical incident reports, both the Assistant Secretary of the Army for Installations, Logistics, and the Environment and the Chief of Staff are notified. The desire is to get the report right, take the time necessary to be credible, and avoid putting out information or sounding alarms that later prove to be unfounded. The task is difficult because frequent ACAMS stack alarms are a common problem; most prove to be false positives rather than chemical events (NRC, 1999a, page 29).

ACTUAL ON-SITE RESPONSES

December 3-5, 2000, Event at JACADS

After the event at JACADS on December 5, 2000, a six-person investigation team was convened, with members from PMCD, the USACAP, and the U.S. Army Pacific, as well as two consultants. The team assembled on Johnston Island on December 13. This team reported its findings on March 15, 2001 (U.S. Army, 2001f). In addition, EPA conducted an investigation on December 7-8, 2000, and issued a report on May 9, 2001 (EPA, 2001). The Army investigation team's agenda was to determine the cause of the event, while the EPA team's aim was to determine whether or not Resource Conservation and Recovery Act (RCRA) violations had occurred. The description below relies heavily on the investigations' written reports (U.S. Army, 2001f; EPA, 2001).

The chemical event report submitted within 3 hours of the event (Appendix A of U.S. Army, 2001f) is necessarily a truncated version of what happened and, as a result, relates a sequence of events that is easily misinterpreted. It states, "At 0156 (local) 05DEC00, a routine sample of ash from the Decontamination Furnace (DFS) was analyzed in the site laboratory that produced levels of VX nerve agent at approximately 2000 [7000]² times greater than Drinking Water Levels (DWL)³ 40,000 ppb vs. 20 ppb." The report also states, "Upon agent detection, the HDC (heated discharge conveyor) bin was moved to the Unpack Area (UPA) and placed under engineering controls." It is important to note, however, that this analysis was for a sample taken from the bin 1½ days earlier (the site's operating procedures allow up to 4 days for samples to be analyzed). A second sample, taken at 0300 local time, was analyzed and reported at 0430 local time, and only then was action contemplated, though not yet taken. Indeed, although the chemical event report states that the bin was placed under engineering controls as soon as the analysis was reported (at 0220 local time), it also notes that the bin was outside engineering controls until 0800 local time.

The Army investigation report (U.S. Army, 2001f) also seems to minimize the importance of the time the event began. It begins: "I. *Introduction*. On 5 December 2000 at 0156 hours (local time), chemical agent VX was detected and confirmed in the ash from the HDC bin (BIN 135) at the Deactivation Furnace System (DFS)."

There is no mention in that report, either in the introduction or in the Executive Summary, of the sample having been taken on December 3. The first mention of the earlier sample occurs on Page 6, under "V. *Event Description*." The report then describes several attempts to analyze the sample on December 4, the suspicion of a false positive, and a request for a second sample.

Had the first sample been analyzed promptly and the results believed, the release of agent to the environment and any potential for harm could have been minimized. This incident illustrates a flaw in the reporting system, which is focused on formal declaration of an incident as a chemical event. The first indication of a problem was an analysis showing VX at approximately 3000 times WCL at 0156 on December 5, but this is not when the "event" was defined as having begun. The question of when a chemical event begins is important because it is the moment beyond which workers, the public, and/or the environment are potentially in harm's way. It also determines the timing for fulfilling the various reporting requirements. It is debatable at what point the evidence was sufficient to declare this JACADS incident an "event," but the *potential* for harm certainly began at 0806 on December 3, 2000, when Bin 135 was removed from the bin enclosure. The most generous interpretation is that event onset began when the site alarm sounded at 1020 on December 5, 2000. Even given this time of onset, the external reporting was tardy. In fact, the event was not reported to EPA Region IX until the compliance officer serendipitously called at 0930 on December 6, 2000, about another matter and was informed of ongoing events. A notice of violation was subsequently issued by EPA on May 9, 2001 (EPA, 2001). Internally, there were indications of notification problems as well; the notification list indicates "1039 completed call-down list." However, several lines were "busy" or resulted in "no answer" or, in one case, "machine."

The subtitle of the Army investigative report (U.S. Army, 2001f), *Report of the 3 December 2000 Chemical Agent **Reading** [emphasis added] in the Heated Discharge Conveyor (HDC) Bin* rather than *Report of the 3 December 2000 Chemical Agent **Event** [emphasis added] in the Heated Discharge Conveyor (HDC) Bin* appears to suggest a continued state of denial.

---

²The bracketed number is in the original document, perhaps indicating confusion about what the actual handwritten entry said.

³DWL is the agent waste protection limit used to assess contamination.

From the December 2000 event at JACADS, it appears that an "event" is assumed to begin when personnel confirm agent release, as opposed to when a release may have actually occurred. The time of onset of an event needs to be clarified.

The problem in defining an event (both whether one has occurred and the date/time of onset) also lies partially in the tendency of the chemical demilitarization personnel to disregard initial indications due to frequent "false positive" readings, as discussed in Chapter 2. The required detection sensitivities test the limits of the technology and lead to many readings that are not verified by subsequent analysis. Modifications, such as ACAMS employing at least two different chromatographic columns, could reduce the number of unverified alarms (false positives). Alternative methods, potentially capable of greater specificity and/or sensitivity, have been suggested in other reports (NRC, 1994).

At the sites where the committee visited there does not seem to be a call-forwarding mechanism for getting information directly to people or a hot line dedicated to notification that an event has occurred. This problem would be amplified at sites where officers to be notified are not in the immediate vicinity.

May 8-9, 2000, Event at TOCDF

After the detection of GB in the common stack at TOCDF on the night of May 8-9, 2000, an investigation was undertaken by a 10-person team, which included representatives from the U.S. Army Nuclear and Chemical Agency, the U.S. Army Center for Explosives Safety, PMCD, the Deseret Chemical Depot, and General Physics Corporation, with partnering from two Centers for Disease Control and Prevention (CDC) scientists. The team completed its information gathering on May 18, 2000, and its report on June 6, 2000 (U.S. Army, 2000b). Separate reports were issued by the CDC's National Center for Environmental Health (May 18, 2000; CDC, 2000), the Utah Department of Environmental Quality (DEQ) (Utah DEQ, 2000a), and the contractor, EG&G (June 16, 2000; EG&G, 2000).

These reports are extensive in their detail, with multiple findings and recommendations, and many addenda. The Army report lists 25 separate findings, 29 recommendations, and four "observations." The CDC report lists 11 conclusions and 15 recommendations. The Utah DEQ report lists eight "concerns," while the EG&G report lists several "direct causes," "root causes," "contributing causes," 11 "findings," and 22 "corrective actions."

Observations

A number of observations can be made from a review of the reports relating to both the December 2000 event at JACADS and the May 2000 event at TOCDF:

1. The various agencies responsible for reviewing incidents took their task very seriously. They made a determined effort to understand the causes of the incident and to recommend changes that would prevent its recurrence.
2. The multiplicity of reports is an example of overlapping investigations that create the potential for lost time for the mission of the program. It is also an indication of communication problems within the chemical demilitarization program. (This observation is elaborated below in this chapter.)
3. Incidents such as the May 8-9, 2000, stack release at TOCDF need to be rare occurrences for such in-depth investigations to be feasible. More frequent investigations of this type would quickly demand more resources than could be made available.
4. The extensive investigation of the May 8-9, 2000, TOCDF incident as opposed to the comparatively cursory examination of the December 3-5, 2000, JACADS incident may be partially attributable to the fact that JACADS was in a shutdown mode while TOCDF will continue operations for several more years. Yet dismantling a plant is not inherently less hazardous than operating a plant. The "waste" mentality (discounting the potential for "mere waste" to result in release of agent) that may have contributed to the JACADS incident needs to be changed, just as does the "crying-wolf too often" mind-set that results from the frequent occurrence of and the use of the term "false positives."
5. It remains to be seen if all of the recommendations in the various investigation reports are actually implemented. Incorporation of such recommendations into the programmatic lessons learned (PLL) program (see Chapter 4) and their subsequent utilization at TOCDF and other sites are necessary responses, if the reports are to be effective.

Following the May 8-9, 2000, event, the TOCDF facility was shut down pending the completion of the various investigations. According to Occurrence Report No. 00-05-08-A1 *Confirmed GB Agent Readings in the Common Stack* (EG&G, 2000), 22 corrective actions were assigned to various individuals on June 19, 2000, at the conclusion of the investigative reports. According to the *Annual Status Report on the Disposal of Chemical Weapons and Materiel for Fiscal Year 2000* (U.S. Army, 2000a), authorizations for operation of the liquid incinerator and metal parts furnace were issued on July 28, 2000, and for the deactivation furnace system on September 21, 2000. Thus, the event led to an approximately $4^1/_2$-month shutdown. It is difficult to assign the exact amount of time for the investigative, corrective, and approval phases needed to commence facility restart because of considerable overlap in phases; i.e., corrective measures and equipment ordering were already occurring as the investigations proceeded.

EXTERNAL AND REGULATORY RESPONSES TO CHEMICAL EVENTS

Applicable Statutes, Regulations, and Guidelines

The activities of the facilities located at the Johnston Island and Tooele sites were governed by multiple statutes and regulatory rules and procedures, as well as permitting requirements. The controlling federal statute, the Resource Conservation and Recovery Act (RCRA; 42 U.S.C. §6901 et seq.), was enacted in 1976. RCRA contains stringent statutory requirements that control the handling and disposal of hazardous waste. The legislation is commonly referred to as the "cradle-to-grave" regulatory procedure and gives EPA's administrator the responsibility to oversee the generation, transportation, treatment, storage, and disposal of hazardous waste. The program can be delegated to the various states for primary enforcement of the statute, although EPA continues to have a federal role of oversight of any such facilities.

Additional statutes that must be considered include the Toxic Substance Control Act (TSCA; 15 U.S.C. §2601 et seq.), the Emergency Planning and Community Right to Know Act (EPCRTKA; 42 U.S.C. §11001 et seq.), the Clean Air Act (CAA; 42 U.S.C. §7401 et seq.), the Chemical Safety Information, Site Security and Fuels Regulatory Relief Act (P.L. 106-40), the Occupational Health and Safety Act (OSHA; 29 U.S.C. 1920.120 et seq.), and the Clean Water Act (CWA; 33 U.S.C. §1251 et seq.), in addition to any state statutes, regulations, and local ordinances. Additionally, as mentioned above and in Chapter 2, the chemical demilitarization program is subject to U.S. Army regulations and specific-site regulations, or standing orders, implemented by the post commander and/or the civilian plant manager. Finally, site activities may also be subject to requirements set forth in memoranda of understanding (MOUs) entered into by government entities and the facility. The MOUs are unique to the site and can address issues specific to the surrounding area and nearby communities.

In addition to national, state, and local regulatory review, there is also oversight required pursuant to the Chemical Weapons Convention (CWC). International CWC observers, commonly referred to as the Inspectorate, maintain offices on site at JACADS and TOCDF. The Inspectorate is responsible for general oversight and for ensuring that the destruction of chemicals is carried out pursuant to CWC guidelines.

These statutes, regulations, and guidelines require notification of outside agencies when incidents affect public health, when permits require such notification, or for the marshalling of assistance in the event of a catastrophe. A review of these international treaties, statutes, rules, and regulations makes it clear that the facilities for chemical demilitarization are highly regulated and can be subject to microscopic oversight. This panoply of regulations befits the extremely hazardous materials that are destroyed by on-site incinerators. Failure to follow the protocols called for by the statutory framework can result in facility shutdowns by the agencies that possess the authority to do so, by court orders, and by the U.S. Army. These failures can also erode public trust. Enforcement of the statutes and regulations can result in notices of violation for failing to operate within a given permit or any number of multiple permits, or for failing to follow reporting procedures. Ultimately there is authority to impose remedial activity sanctions, civil fines, and in the worst case, criminal fines and imprisonment.

Following a serious chemical event, it is typical that there is an investigation that can originate from multiple state and federal regulatory agencies. For instance, the state environmental agency may assume the lead investigative position, although the EPA always retains the authority to initiate its own independent investigation.

Time requirements for verbal reporting and follow-up written reports are not unique to chemical demilitarization facilities. Furthermore, the regulatory process is not static—it evolves. The same is true for the permitting process. Renewals are a part of the process, with a period of time built in prior to the expiration of permits. This provides the regulated community with an opportunity to revisit and implement technological advances by the operating unit. The trend to tighten the regulation to a higher standard of compliance affects all regulated facilities.

Each facility develops a regulatory history with the enforcement agencies with which it works. Candor and trust are essential for these relationships to succeed. Failure to follow incident reporting procedures, as agreed upon in advance of an incident, erodes trust that is critical to chemical demilitarization operations, wherever they are located. The facilities begin operations under a cloud of suspicion, often due to public misunderstanding, lack of public education and information, media hyperbole, and general "NIMBY" (not in my back yard) sentiments. Poor communication with the regulatory agencies and the public will further erode the program's public involvement and regulatory agency trust (NRC, 2000b).

Memorandum of Understanding Between Deseret Chemical Depot and Tooele County

In the case of TOCDF, because of and subsequent to the May 8-9, 2000, incident, Tooele County entered into an MOU (Utah DEQ, 2000b) in September 2000 (updated in November 2001) with the facility that (1) defines specific event classifications; (2) identifies and displays hazard predictions for chemical operations with a potential for producing agent effects beyond the installation boundary; (3) provides recommendations for protective actions to be taken in advance of potential events; and (4) conducts daily activities that will mimic and reinforce emergency activities, thereby enhancing the notification and response abilities of Deseret Chemical Depot (DCD) and Tooele County. Thus, the facility and Tooele County have an agreed-to daily protocol concerning the tasks that will be undertaken on a particular day, the times and type of agent munitions that will be processed, and the meteorological data that will be obtained during each operation. Under the MOU, Tooele County is required to inform the DCD of any special events, projects, or other activities occurring in the community that could affect a quick and safe evacuation of DCD. Examples given were special events drawing unusually large crowds, road construction, bridge work, and so on. In the event of a chemical incident, Tooele County must inform DCD and Utah Comprehensive Emergency Management of the protective action decisions they have made (see Appendix G).

The parties agreed to the following terms for classifying emergency events:

- Routine leaker or agent detection within containment
- Non-surety event
- Limited-area event
- Post-only event
- Community event.

Definitions for each of these classifications, as well as the body of the MOU, are reprinted in Appendix G.

For the last three of the five event categories listed above, DCD has agreed that notification shall be made to Tooele County within *10 minutes* of when chemical agent is detected in the atmosphere, i.e., outside engineering controls, and when other unusual circumstances occur, even if a chemical event is only suspected. DCD also agrees to use the dedicated "Chemical Notification Hotline"[4] telephone as the primary means of notification for routine leakers and other occurrences of chemical agent detection outlined above, as well as for events falling into the defined chemical event classifications (Utah DEQ, 2000b).

Had the above terms of notification and procedures now specified in the MOU been in place at the time of the May 8-9, 2000, incident at TOCDF, the impermissible delays between the time of detection and the time of reporting could have been avoided. The MOU between DCD and Tooele County and the new reporting procedures address a number of the recurring reporting deficiencies that have been experienced at the site. Missing from the MOU, however, are specific training requirements that should be implemented to ensure that the proposed reporting system can be implemented effectively.

Levels of Investigation

The multiple investigations of the May 8-9, 2000, Tooele chemical event probably prolonged operational shutdown unnecessarily. Arguably, multiple levels of review by independent agencies increase the ability to thoroughly characterize an incident. There is a point, however, where the scale tips and accuracy and completeness give way to redundancy and inefficiency with no added benefits.

The loss of operating time is expensive. During the committee's visit to TOCDF, operating staff estimated that the cost to operate the Tooele chemical demilitarization facility is approximately $10,000 per hour or $240,000 per day (U.S. Army, 2001g). Long facility shutdowns also lead to a deterioration of operating skills. Facility down-time following chemical events can be minimized by implementing policies that permit a coordinated review effort between multiple oversight entities, in addition to the development and submittal of a single comprehensive incident report. Preagreement among responsible oversight agencies to establish a single review team with a predetermined distribution of representatives from various agencies and their areas of expertise would allow the rapid deployment of a single, comprehensive event investigation.

Consolidating the investigation process can still ensure that the facilities are operating with the highest margin of safety, while at the same time ensuring that procedures are in place that will minimize plant shutdown time following chemical events or other safety infractions.

MODELING POTENTIAL POPULATION EXPOSURE

When chemical agents are released into the atmosphere, a key challenge is to predict the affected population's exposure. This information is needed for developing effective evacuation plans and implementing any needed mitigation measures. Figure 3-1 illustrates the four elements that must be integrated, the linkage between these components, and some of the information needed to perform the calculations.

As used by the Community Stockpile Emergency Preparedness Program (CSEPP), the current implementation of the system shown in Figure 3-1 is called D2PC, which is used to calculate dosages and concentrations from accidental releases of chemical warfare agents. The model is based on a Gaussian plume/puff formulation for transport and dispersion in the atmosphere (Seinfeld and Pandis, 1998). D2PC is a revision of an older dispersion model, D2, which was documented in 1982. The D2PC model runs on a personal computer and is based on the technical paper "Methodology for Chemical Hazard Prediction" (DoD, 1980). The June 1992 revision of D2PC was the version originally approved by the Army for use by the CSEPP. Subsequently, D2PC has under-

---

[4]The Chemical Notification Hotline is a dedicated phone line between DCD and Tooele County. The Chemical Notification Form (see Attachment A of Utah DEQ, 2000b) provides the format for any information communicated via the Hotline.

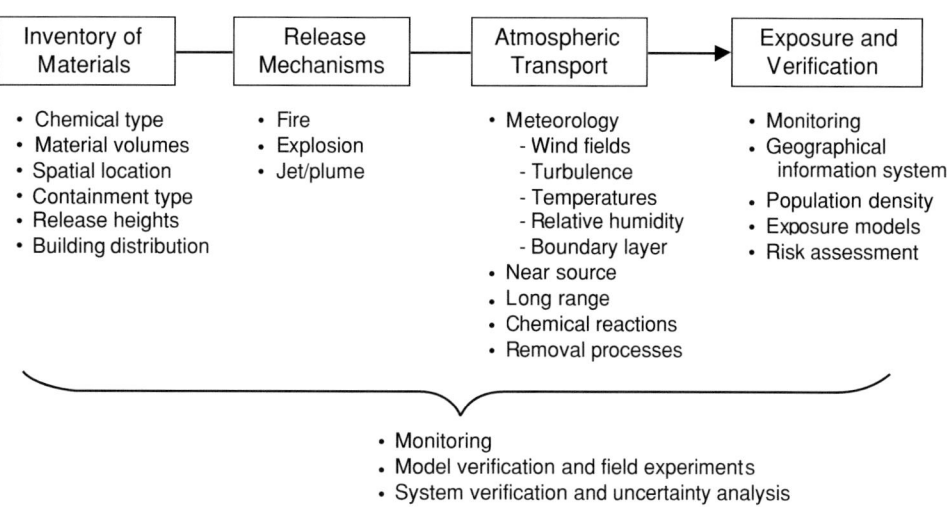

FIGURE 3-1 Component parts of an integrated system for modeling the impact of release of chemical agents.

gone at least two revisions. In June 1994, the U.S. Army Nuclear and Chemical Agency approved an October 1993 version of D2PC for all CSEPP and chemical stockpile emergency planning and response actions. In March 1997, it approved the Emergency Management Information System (EMIS), version 3.0 (with the exception of the automated calculation of atmospheric stability), for CSEPP as well. D2PC was most recently upgraded in March 1998, and it is this version that is embedded in EMIS 3.1.

The D2PC model is currently supplemented with the Partial Dosage (PARDOS) model, which uses the D2PC methodology to predict cloud arrival and departure times and dosage accumulation times. The D2PC/PARDOS models assume flat terrain and steady-state meteorological conditions. Many demilitarization sites, however, are in regions of complex terrain, and the steady-state assumption is realistic only for small, short-term releases.

Gaussian puff/plume dispersion modeling techniques embedded in D2PC are representative of the state of the art in the late 1970s. Since then, there have been many technical advances in understanding atmospheric turbulence, boundary layer structure, and the effects of complex terrain that could benefit the CSEPP program.

In 1996, in response to some of the limitations of D2PC, the Army tasked Innovative Emergency Management, Inc., to develop a new model called D2-Puff. D2-Puff predicts dosages and concentrations in changing meteorological conditions, including wind shifts. D2-Puff uses the same methodology for release of agents and the same atmospheric dispersion coefficients as D2PC. The technical basis for the model and its verification are described in three comprehensive documents (IEM, 2001a,b; U.S. Army, 1999b). At present, the modeling system is used in two modes. In the first, a planning mode, the model is used to determine potential population exposure to agent at a particular level in accident scenarios that might occur during routine operations. In the second mode, when emergencies occur the system is used to predict the dispersion of the agents and the likely population exposure. D2-Puff includes the following new features and capabilities:

- A Lagrangian puff model that allows concentrations and dosages to be calculated when meteorological conditions change in time or vary over a region
- The calculation of concentrations and dosages within enclosed structures, such as buildings used as shelters
- The ability to handle multiple release locations
- The ability to simulate dosages received by individuals who are exposed to only a portion of a plume
- The ability to include meteorological observations from multiple locations
- The ability to include data from weather forecasting models (assuming that a suitable meteorological data assimilation capability is attached to D2-Puff)
- The ability to model the effects of complex terrain on plume motion
- The ability to compute dispersion based on measurements of the variance of wind direction
- The ability to compute for acute exposure guideline levels (AEGLs) (NRC, 2001b)[5]
- A graphical user interface.

---

[5]Acute Exposure Guideline Levels (AEGLs) are a hazard communication measure developed by the National Advisory Committee on Acute Exposure Guideline Levels for Hazardous Substances. The committee developed detailed guidelines for devising uniform, meaningful emergency response standards for the general public. The guidelines define three tiers of AEGLs as follows:

Another important change in approach concerns the way that the hazard is represented in D2-Puff. D2PC produces cigar-shaped footprints for 1 percent lethality, no-deaths, and no-effects dosages. With D2-Puff the analyst can no longer think of dosages solely in terms of distances or relatively simple cigar-shaped footprints. With varying meteorological conditions, D2-Puff produces irregularly shaped footprints for 1 percent lethality, no-deaths, and no-effects dosages. The no-effects footprint for D2-Puff will generally not be as long as the no-effects footprint (no-effects distance) for D2PC, although the D2-Puff footprint will generally be wider. This difference will have an impact on protective action decisions. As with D2PC, D2-Puff indicates that persons living in the downwind direction near a release will be the first exposed to the hazard. However, the wind direction may shift before populations farther away are exposed to the hazard. This wind shift may result in exposure of a broader area in the immediate vicinity of the release location—an area larger than the initial downwind path of the plume. In this situation, emergency managers may find that they have to change their priorities for protective actions.

The D2-Puff model, and other plume dispersion models, can be calibrated for the effects of complex terrain at specific sites by experimental releases and downwind measurements of an inert gaseous tracer under a variety of representative meteorological conditions. These calibrations can significantly enhance the accuracy of dispersion calculations from specific fixed sites like chemical agent storage yards and demilitarization facilities.

While D2-Puff represents an advance in capabilities over D2PC, it is still based on Gaussian dispersion modeling with its attendant limitations. Perhaps the most serious limitation of the D2-Puff/D2PC methodology for chemical hazard prediction arises from the neglect of the variation in wind speed with height. Because both the D2-Puff and D2PC models assume that the wind speed measured at 10 m above ground level is representative of the transport wind speed at all downwind distances, they tend to overestimate transport speeds for low-level releases at short range and underestimate transport wind speeds for all release heights at longer downwind distances. Thus, the toxic cloud produced by a large accident will arrive in areas more than 1 to 2 km from the release sooner than predicted by the models. This is especially relevant to sites close to population centers. A further limitation of the Gaussian dispersion formulation is its low predictive accuracy for long-range transport (>50 km). If a substantial release were to occur, the current D2PC/D2-Puff models are not suited for predicting the impacts on populations that might be 100 km or more downwind from the release site. As with any model, the results produced are limited by the accuracy of the inputs. These limitations include uncertainties about the amounts of chemical agents released and about meteorological conditions. D2-Puff, like other models, can produce hazard estimates that are helpful for emergency planning and response.

In light of the limitations of Gaussian dispersion models, a key part of the CSEPP should be an ongoing evaluation of alternative approaches to modeling the release and impact of chemical agents. A considerable wealth of relevant modeling experience has been developed for coping with such events as fires and explosions at chemical plants, transportation spills, nuclear accidents, tunnel fires, uncontrolled forest burns, volcanic eruptions, and oil well fires. Many different models and methodologies are available. For example, one option would be to supplement each stockpile site with the capabilities of the National Atmospheric Release Advisory Center (NARAC)[6] that is located at the University of California's Lawrence Livermore National Laboratory.

A more accurate modeling capability is valuable only if it is coupled with timely communication of results and appropriate responses by the stockpile site and surrounding communities. In the case of sites located close to large communities it is particularly important to have fast communication and alert procedures. The committee found, based on several site visits and interviews, that these procedures should be reviewed to identify bottlenecks that could be removed through better communications technologies.

EMERGENCY RESPONSE: PREPAREDNESS, PLANS, NOTIFICATION, AND COORDINATION AT TOCDF

This section focuses on the May 8-9, 2000, TOCDF incident but also draws on the December 3-5, 2000, JACADS event in discussing the importance of reporting requirements. The TOCDF incident is the primary focus because of that

---

AEGL-1: The airborne concentration of a substance above which it is predicted that the general population, including susceptible individuals, could experience notable discomfort, irritation, or certain asymptomatic nonsensory effects. However, the effects are not disabling and are transient and reversible upon cessation of exposure.

AEGL-2: The airborne concentration of a substance above which it is predicted that the general population, including susceptible individuals, could experience irreversible or other serious, long-lasting adverse health effects or an impaired ability to escape.

AEGL-3: The airborne concentration of a substance above which it is predicted that the general population, including susceptible individuals, could experience life-threatening health effects or death.

Guidelines for each of the three levels of AEGL—AEGL-1, AEGL-2, and AEGL-3—have been developed for each of five exposure periods: 10 minutes, 30 minutes, 1 hour, 4 hours, and 8 hours. See NRC (2001b).

---

[6]NARAC is a national emergency response service for real-time assessment of incidents involving nuclear, chemical, biological, or natural hazardous material. NARAC's primary function is to support the Department of Energy and the Department of Defense for radiological releases. Under the auspices of the Federal Radiological Emergency Response Plan and the Federal Response Plan, the state-of-the-art NARAC modeling system has the capability to perform assessments of impacts from local to global scales. More information is available online at <http://narac.llnl.gov>.

event's potential implications for the safety not only of the workers at the plant, but also for residents in the nearby community. This National Research Council (NRC) committee is not the first to express concern about the emergency response and management capabilities at TOCDF. Previous findings and concerns regarding the response system noted by other NRC committees and the General Accounting Office (GAO) (see Box 3-1) provide some necessary context for the committee's examination.

This committee's evaluation of the emergency response to the two JACADS and TOCDF incidents that it examined in detail focuses on how effectively the division of responsibilities between the Army and the Federal Emergency Management Agency (FEMA) (see Box 3-1) actually functioned, and analyzes how it is likely to continue to function in the future. Although it is critical to have well-exercised plans, a communication system that enables adequate warning, effective communication among responders, and personnel who are appropriately attired for the nature of the hazard, it is equally critical that the organizational structure functions as designed, enabling an effective response. Indeed, how effectively the emergency response system is organized and how capable it is of functioning in a coordinated fashion have important implications for the three additional incinerator-based chemical demilitarization sites that are close to beginning operations. One of the important components of this committee's examination of the emergency response to the two JACADS and TOCDF incidents has been a review of the preparedness of the emergency management system when required to function during stressing events.

Relevant to the examination of emergency preparedness are a recent GAO report that examined FEMA's and the Army's efforts to prepare states for chemical weapons emergencies (GAO, 2001) and a CSEPP report describing CSEPP and Army benchmarking of the system (CSEPP, 2000). As pointed out in the GAO report, FEMA has adopted a series of national quantitative performance indicators that use benchmarks to evaluate the preparedness of different states in the program (GAO, 2001). These benchmarks are supposed to focus on outcomes rather than outputs as measures of performance in ensuring the essentials of public safety, including warning system effectiveness, readiness of coordination systems, reliability of critical communication systems, and public awareness of protective actions. FEMA is responsible for benchmarking emergency management compliance off-post; the Army uses a similar system at its installations (GAO, 2001). The 2001 GAO report also mentions that Utah is one of three states considered to be fully prepared for a chemical emergency and that an active cooperative effort by the community is essential to the state's current state of preparedness. Interestingly, these three states are considered by FEMA and the Army to be fully prepared, even though both the Army and FEMA have failed to issue any site-specific planning guidance for local communities or states covering reentry into a contaminated area of a com-

---

BOX 3-1 Previous Concerns About and Recommendations for Achieving Efficient CSEPP Operations

In its first systemization report produced when the plant was about to begin operations in 1996 (NRC, 1996), and as summarized in the National Research Council (NRC) report Tooele Chemical Agent Disposal Facility—Update on National Research Council Recommendations (NRC, 1999a), the NRC's Stockpile Committee called on the Army and the Federal Emergency Management Agency (FEMA) where appropriate to:

1. ensure that local and state Chemical Stockpile Emergency Preparedness Program (CSEPP) plans for responding to chemical events were complete and well exercised
2. increase its efforts to work with the Utah Division of Comprehensive Emergency Management to ensure that first responders were adequately trained to use personnel protective equipment
3. make certain that the Army/FEMA provided the necessary resources for completing the planned Tooele County emergency communications system.

In 1999, the NRC added another recommendation: that the Army ensure that CSEPP and FEMA officials understand how the quantitative risk assessment (QRA) and other activities might affect risk and reflect this understanding in emergency planning and preparedness activities (NRC, 1999a). The 1999 NRC report reviewed and updated recommendations on operations at the Tooele Chemical Agent Disposal Facility (TOCDF). It noted that in accordance with the formal reorganization of responsibilities that had just been carried out between the Army and FEMA, all on-site responsibilities for emergency management were retained by the Army and all off-site responsibilities for emergency management and planning were given to FEMA. The 1999 Stockpile Committee report, noting previous General Accounting Office (GAO) reports that had cited existing problems with the CSEPP, stated, "The Committee is also concerned about CSEPP and about the horizontal fragmentation of responsibility at the federal level." The report further commented (NRC, 1999a):

> Previous briefings by directors (both Army and FEMA) of the CSEPP, as well as discussions with directors of state emergency management agencies, have all stressed the importance of a well-coordinated response-management capability.... The recent reorganization will require excellent coordination and communication to overcome the barriers of separate organizational responsibilities.

Finally, the 1999 NRC report expressed skepticism about the reorganization's impact on improving the capacity for responding to an emergency.

munity, or guidance on when it is appropriate to notify citizens to leave shelters following an event.

The committee judges that the benchmarks demonstrate a significant effort by FEMA and the Army to coordinate their efforts to measure a program's status and to guide funding. For example, these measures have been developed over time and include the initial guidance document issued in 1993 (FEMA, 1993), and revised in 1996 (FEMA, 1996) to include nine benchmarks. These benchmarks were later revised again in 1997, and then again in a joint policy paper (FEMA, 1997) that added three additional benchmarks. The GAO used these 12 agreed-upon benchmarks in 19 "critical items" for its review of the program.

The development of jointly used benchmarks does not reveal the full extent of the efforts by the Army and FEMA to jointly coordinate the emergency response/management system for chemical incidents. On October 8, 1997, coinciding with the formal division of the program, an MOU between the Army and FEMA formally identified their respective roles and responsibilities and joint efforts for " . . . emergency response, preparedness involving the storage and ultimate disposal of the U.S. stockpile of chemical warfare material" (FEMA, 1997). Despite these efforts, the GAO has continued to find uneven performance measures being used and a lack of effectiveness in providing technical assistance and guidance to the states and communities (GAO, 2001).

The performance of the emergency management system during the TOCDF May 8-9, 2000, event is not reassuring. It raises questions about how to interpret the system's performance and what is meant by the term "fully prepared." The lack of timely notification that an event had occurred has several important implications. First, benchmarking performance evaluations aside, the real test of an emergency management and response system is how it functions during an incident rather than performance during training exercises. What is particularly troubling is that something as simple as notification of an alarm (even after it was confirmed) was not reported to the Tooele County Emergency Operating Center (EOC). No one disputes the fact that the Tooele County EOC and Utah officials should have been notified of the events. This notification is part of the standard operating procedures (SOPs) and is probably the most exercised component of the system during operations testing and exercises. The fact that SOPs were clearly disregarded, and the off-site community potentially put at risk because of the lack of notification and knowledge of the event, demonstrates a clear breakdown of the system at the most elementary level. While some action aimed at preventing the repeating of this sequence of events has been taken through a new MOU for Information Exchange (Utah DEQ, 2000b), as discussed previously, the events surrounding this incident raise questions in critics of the program concerning the trustworthiness of those in charge of the emergency response and notification system. This trust is crucial to surrounding communities' participation and cooperation in these programs, and questions concerning the credibility and functioning of the emergency notification and response systems have serious implications not only for communities where operating systems are currently located, but also for communities where they are planned, like Anniston. As pointed out above, this cooperation was cited in the GAO 2001 report as being a fundamental condition for the three programs gaining fully prepared status from FEMA and the Army.

During their tours of TOCDF and DCD, members of the committee raised questions concerning the responsibilities of personnel as they related to the Tooele County EOC and Utah DEQ. In several instances personnel reported that their responsibilities "ended at the fence" and that they were not responsible for emergency management operations in the community. Similar attitudes were expressed at the JACADS facility, although the lack of a community near the facility mitigates the impact of such views. Technically, this view is correct concerning the division of responsibilities. However, for an effective response the program requires a strong degree of coordination between the DCD EOC and the Tooele County Office of Emergency Management.

It should be remembered that at both JACADS and TOCDF the emergency response system functioned with only a few problems (such as those at JACADS when important personnel could not be notified because of communication problems). That is, the failures of notification occurred in alerting the civilian authorities that are a part of CSEPP. Within the Army structure at JACADS, for example, the personnel were assembled at checkpoint "Charlie" for possible evacuation once the alarm was sounded. The plant control room at Tooele informed the DCD's EOC in a timely fashion of the alarms and provided it with updates on the situation. However, the DCD EOC then failed to pass on the notice to the Tooele County EOC and relevant State of Utah agencies. It is impossible to determine how the CSEPP portion of the emergency management system functioned as it was not provided timely notification of the events. Other communities soon to host chemical demilitarization facilities can learn a good deal from these two events and the nature of the "fix" that has been made by the Army and Tooele officials. Given this failure of communication and adequate notification, it is reasonable to assume that efforts to correct the problems associated with the response would focus on information exchange, such as through the MOU entered into by Tooele County and the DCD (Appendix G).

The lack of notification and warning between the DCD and Tooele County and appropriate local and state agencies was caused in part by a lack of coordination between components of the two programs (CSEPP/FEMA and the Army), and in part because of DCD's emergency management responsibilities that "end at the fence" (although timely communication cannot). The recent GAO report (GAO, 2001) on FEMA and Army efforts to prepare communities for a chemical emergency is vague on how to improve what is being done other than suggesting that the two entities become proactive in doing so.

Even if the various components of the emergency response system are designed to be fully coordinated, the system will not function well unless there is a high level of trust among the personnel involved. In particular, there needs to be trust between those "inside the fence" (professional personnel) and those "outside the fence" (local officials and the public).

PUBLIC RESPONSES TO CHEMICAL EVENTS

A significant aspect of the responses to chemical events concerns when and how the event is communicated to local officials and the local public.[7] While much of the focus of post-event response is necessarily on the requirements of the formal regulatory process, interactions with the affected local officials and public have important implications as well. From the perspectives of the public and their officials, "chemical events" are largely involuntary risks that are potentially catastrophic and of technological origin. These characteristics render chemical events and incidents subject to substantial "social amplification" in which the characteristics of the events interact with individuals' perceptions of the risk associated with them and the pattern of communication with the public and their response to both the event and the communication (Kasperson, 1992; Kasperson et al., 1988).

According to this formulation, news reporters, interest groups, and concerned citizens monitor events and select and retransmit risk signals pertinent to those events via the news media and informal networks, which in turn results in a ripple effect of secondary impacts. These secondary impacts could include changes in perceived levels of risk, altered trust for the organizations and officials involved, pressure for legal and institutional change, changes in property values, and a myriad of other effects. Thus, the pattern of communication with and responsiveness to the public and their officials can have substantial "real" effects beyond the immediate health and environmental impacts posed by the chemical event. From a programmatic perspective, most importantly, these secondary effects can delay and further debilitate a program by undercutting the credibility of the agency(ies) entrusted with implementing the program, reinforcing negative messages about the technology being utilized and leading the public to question reports and official statements about progress in meeting program objectives.

Understanding how chemical events might initiate the "social amplification" process is facilitated by elucidating critical aspects of the trust relationship engendered by activities such as the chemical weapons demilitarization program. Officials and citizens of the affected local communities, along with national officials, share the objective of destroying the chemical stockpile but must rely on others to carry out that destruction in a safe and timely manner. To undertake the program, these "principals" must establish a relationship with agencies (PMCD) and contractors—or "*agents*"[8]—to carry out the mission.[9] The technological requirements of the process, and the magnitude of the potential hazards, lead to barriers of complexity and security that—for practical purposes—make the program difficult for the principals to directly evaluate and monitor. The theory of principals and *agents* is discussed further in Chapter 1.

Effective management of the principal-*agent* relationship in the chemical demilitarization program in order to achieve the required level of trust appears to require (1) monitoring processes that assure principals of their role in effective oversight, (2) complete and timely disclosure of events by the *agents*, and (3) demonstrable and timely assessments of the problems leading to chemical events and their correction.

The JACADS December 3-5, 2000, incident raises several important issues concerning interactions with external principals. First, failure to believe the first sample analysis and act immediately to isolate the contaminated material is troubling. Absent very careful monitoring (in the form of investigations) by regulators, the event would have been misunderstood, potentially inhibiting appropriate responses. Second, tardy compliance with reporting requirements (as discussed above in this chapter), even when very permissive assumptions are made about the timing of the event onset, may well raise significant concerns among public officials, media, and affected citizens. Though JACADS is itself a geographically isolated facility, if the lapses associated with the December 2000 incident are repeated at other sites, residents living near similar facilities might lose confidence in the monitoring process. Moreover, these incidents could be seen as indicators of larger, unobserved problems in plant operations, such as insufficient willingness to forthrightly identify and correct conditions that could lead to chemical events.

The committee's investigation did not indicate that JACADS personnel intended to distort the December 3-5, 2000, event or delay reporting. However, the context (the "mere waste" mind-set versus the "agent" mind-set) and outcome could erode the confidence of external principals at a continental U.S. site in the monitoring and control processes.

---

[7]Almost by definition, the communication process includes the local news media and interested citizens' groups.

[8]It is unfortunate that use of the term "agents" to indicate those who carry out tasks for "principals" might in this report be a source of confusion in the context of the chemical demilitarization program (where "agent" usually refers to chemical agent). Where *agent* is used in the institutional sense, it is italicized to reduce the potential for confusion.

[9]There is a large and growing literature on what is referred to as the principal-*agent* relationship. For some of the more important work, see Wood (1992) and Scholz and Wei (1986).

The substantial costs in terms of resources and time required for multiple investigations of chemical events involving environmental releases, such as those that occurred in the TOCDF May 8-9, 2000, event, might contribute to a defensive mentality on the part of the operating personnel. At the same time, it is essential that local officials and local citizens have trusted representatives involved in these investigations both to ensure that they are competently undertaken and to facilitate effective communication of the results. The need for such local representation is underlined by the findings of delayed reporting or failure to report, indicating the significant flaws in the reporting process that stimulated the new notification and communication MOU between the DCD and Tooele County.

4

Implications of Past Chemical Events for Ongoing and Future Chemical Demilitarization Activities

Chapters 1 through 3 of this report are based on an examination of activities at Johnston Atoll Chemical Agent Disposal System (JACADS) and Tooele Chemical Disposal Facility (TOCDF), both of which employ baseline incineration systems to destroy chemical agents. Third-generation incineration facilities are scheduled to begin operation in 2002 or 2003 at Anniston, Alabama, Umatilla, Oregon, and Pine Bluff, Arkansas. The committee believes that many of the observations and recommendations made in this report are applicable to all demilitarization facilities, including those that may not use incineration.

Evidence indicates that chemical demilitarization incineration facilities are safe as designed if they are operated properly and if the appropriate operating procedures and protocols are in place (NRC, 1996). The avoidance of risk during any type of process upset depends on having the necessary engineering controls in place and on the operator's skill and training in using them to advantage. This level of preparedness requires in turn that a thorough hazard risk analysis be performed and that all personnel be thoroughly trained and given refresher courses at appropriate intervals. At both JACADS and TOCDF, extensive written procedures are in place for normal operations as well as for startup and shutdown, and operators receive systematic refresher training in these procedures. It can be expected that future chemical demilitarization facilities will also operate this way. Key factors for minimizing—if not eliminating—chemical events in the future include:

- Sound risk and change management programs and procedures;
- Effective safety programs that are focused on continuous improvement, and have the full visible support of all levels of management; and
- Systems for efficient and timely program-wide dissemination of information and communication.

RISK AND MANAGEMENT OF CHANGE PROGRAMS ALREADY IN PLACE

This section describes the procedures that are in place for evaluation of change, including the risk associated with a change. The current Chemical Stockpile Disposal Program (CSDP) risk management program is fully described in *Risk Assessment and Management at Deseret Chemical Depot and the Tooele Chemical Agent Disposal Facility* (NRC, 1997). It is a multilevel program that defines policy, sets requirements, provides guidance on implementation, and, at the facility level, defines specific requirements the facility must meet and specific management processes that must be implemented. The CSDP risk management program is based on a long history of safety and hazard analysis and regulation by the Army. An informal risk management process was developed at the TOCDF in parallel with the site-specific quantitative risk analysis (QRA). This process was described in the NRC report *Review of Systemization of the Tooele Chemical Agent Disposal Facility* (NRC, 1996), which summarized a number of plant and operational changes that had been implemented as a result of accident scenarios identified in preliminary work on the QRA. As part of the risk management process, the following risk-monitoring activities have been introduced:

- Performance evaluation (based on feedback from activities and incidents);
- Emergency response exercises (periodic exercises on site, with Chemical Stockpile Emergency Preparedness Program (CSEPP) personnel);
- Risk tracking (as new data become available, as risk models are improved, and when changes occur in the facility, the related changes in risk related to safety, environmental protection, and emergency preparedness will be calculated and tracked); and

- As required by the Program Manager for Chemical Demilitarization (PMCD) now for essentially all facilities, participation in meetings and/or teleconferences about design lessons learned and programmatic lessons learned.

The Army's formal risk management process is described in a program-wide document, *Chemical Agent Disposal Facility Risk Management Program Requirements* (U.S. Army, 1996c), which provides a basis for the CSDP risk management program. The risk management program is a framework for understanding and controlling all elements of risk within the disposal facility and the stockpile storage area. It links risk management needs to other specific requirements of the Army and other parties at top levels of management and identifies specific documents and references that apply to all CSDP facilities.

In January 1997, the Army issued its draft, *A Guide to Risk Management Policy and Activities (the Guide)* (U.S. Army, 1997b). This draft provides an overview of the processes for managing risks associated with Program Manager for Chemical Demilitarization (PMCD) activities and describes a process for managing changes that may affect the risk associated with PMCD activities. It defines issues that are matters of risk assessment and issues that are matters involving policy (value judgments) and attempts to establish an approach to integrating them and to involving the public in that integration.

The PMCD policy indicates that risk management is integrated into the normal functioning of the organization:

- Operations are now based on the risk management program requirements document (U.S. Army, 1996c).
- The Risk Management and Quality Assurance Office has been assigned the task of integrating risk management for operations, design, and construction.
- The Environmental and Monitoring Office has been assigned the task of assessing hazards to the environment, the populace, and biota in terms of regulatory requirements.
- The CSEPP has the task of planning for potential emergencies and providing liaisons with other emergency preparedness organizations. Note that this program is not a part of PMCD.
- The Public Affairs Office is charged with providing liaisons among the public, the Citizens Advisory Commission (CAC), state authorities, and the Army to facilitate public involvement.

Another significant element in risk management is the management of change. Although changes are presumably made for good reasons, the overall safety of the facility could be compromised if the effects of change on risk estimates are not evaluated or understood. Changes need to be documented and analyzed to determine if they affect procedures, training, or other aspects of the program. This established configuration is based on the initial design of the facility and incorporates changes that have been approved and implemented. The established configuration is the basis for the plant's up-to-date health risk assessment (HRA) and QRA.

If a proposed change is significant, assessing its value is acknowledged to be both a policy question and a factual question. Structured discussions focus attention on all factors that affect the decision, and information on the impact of the proposed change in significant cases should be made available to the public, to the CAC, and to state regulators, and public comments should be solicited when appropriate to the change contemplated. For the most significant changes (Resource Conservation and Recovery Act (RCRA) Class 3) the Army, with the assistance of the controlling regulatory body, must schedule a public hearing. The definition of a RCRA Class 3 change is embodied in the existing federal regulations. The Army's decision will take into account community desires (where appropriate to the complexity of the change) and needs as well as important facts and intangible factors, which are summarized in Table 4-1. Note that factor 6 in Table 4-1, "comparison to previous decisions," ensures either that decisions are consistent or that the reasons for inconsistencies are clearly stated. A thorough consideration of uncertainties is also required. The Army is tasked to prepare

TABLE 4-1 Issues and Factors in Assessing the Value of Change Options

1	Public Input
2	QRA Risk a. All available QRA risk measures, including expected fatalities, cancer incidence, fatalities at a one-in-a-billion probability, and probability of one or more fatalities b. Risk trade-offs: public versus worker, individual versus societal, processing versus storage c. Uncertainties in the technical assessment of risk d. Insights from sensitivity studies
3	Hazard Evaluations
4	HRA Risk a. Insight from sensitivity studies
5	Programmatic a. Cost of the change relative to other proposals and program objectives b. Schedule for implementation c. Uncertainties in estimates d. Impact of implementation on overall objectives and schedule for disposal of the weapons and chemical agent e. Consideration of the improvement anticipated by this change with other proposed improvements
6	Comparison to previous decisions

SOURCE: Reprinted from U.S. Army (1997b), p. 53.

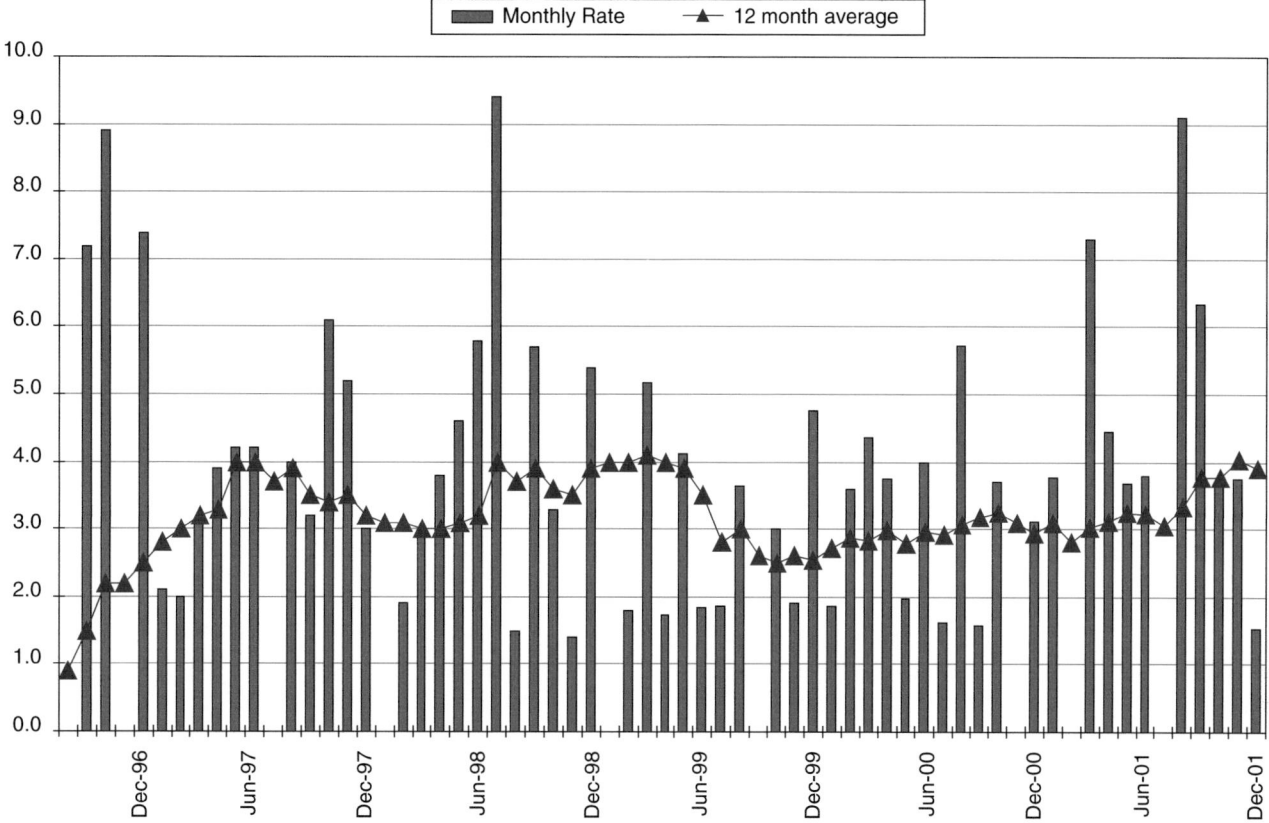

FIGURE 4-1 TOCDF recordable injury rate 12-month rolling average, August 1996 (the start of agent operations) through December 2001. SOURCE: Data up to 1998 from NRC (1999a); data for 1999 to 2001 from U.S. Army (2002b).

responses to all public comments and inform regulators and the CAC of decisions and their rationale.

If all issues are considered in an appropriate and timely manner, general consensus may be possible. But even if consensus is not reached, the Army, as decision maker, will provide a "synopsis of the considerations and a summary of the overall decision basis, listing the rationale for each factor" (U.S. Army, 1997b). In this way, interested parties can see if their concerns were considered and what effect they had on the decision.

SAFETY PROGRAMS

The safety of the public, the environment, and workers is a very significant part of a congressional mandate for the conduct of the chemical demilitarization program. The NRC's Stockpile Committee previously expressed concern over production (agent destruction) having a higher priority than safety—at least from the standpoint of the contractors' award fee criteria (NRC, 1999a and 2002). Responding to this observation, the Army revised the criteria to emphasize safety and production equally. An additional concern expressed repeatedly by the Stockpile Committee is a preoccupation with agent safety, to the detriment of traditional occupational health and safety programs and performance, and it has urged plant management to lead the operating sites toward a "safety culture" (NRC, 1999a).

At JACADS, significant progress was made in developing a safety culture, and during the latter phases of demilitarization operations the plant was consistently achieving excellent safety performance. This does not appear to have been the case at TOCDF (NRC, 1999a).

Although traditional performance indicators such as recordable injury rates (RIRs) at TOCDF are comparable to all-industry averages, there has been very little improvement in these metrics since operations began (Figure 4-1). Nor is there an indication that TOCDF has moved toward a safety culture at any appreciable rate, even though management has developed a TOCDF safety culture plan and has implemented several programs aimed at achieving the safety plan's goals (NRC, 1999a). No additional findings or observations resulted from this study. The Committee on Evaluation of Chemical Events at Army Chemical Agent Disposal Facilities concurs with the Stockpile Committee in observing that the TOCDF is being operated in a safe manner, but that it can and should be continuously improving its safety programs and performance.

The Committee on Evaluation of Chemical Events also concurs with the Stockpile Committee in its belief that future demilitarization facilities should be safer at start-up, as evidenced by performance metrics, than their predecessors. Such performance should not be difficult to achieve, given an effective programmatic lessons learned (PLL) program and the fact that several managers with chemical demilitarization experience will be working at the newer sites. Management and employees at new sites must begin the process of establishing a safety culture before operations commence.

PROGRAMMATIC LESSONS LEARNED PROGRAM

The PLL program—the principal means of communicating lessons learned both within and among the various chemical demilitarization facilities—is the PMCD's only significant vehicle for communicating and coordinating risk, design, and operational issues among sites. The PLL program until recently was administered by PMCD with support from Science Applications International Corporation (SAIC). The program manager at SAIC was hired specifically because of his background in and extensive familiarity with detailed operating procedures, training, and quality control in a hazardous and demanding environment.

Dr. Mario Fiori, Assistant Secretary of the Army for Installations and Environment, presented his vision of some changes in the management and operating philosophy for the chemical demilitarization program to the Stockpile Committee on June 29, 2002. A major thrust of his presentation was that the contractors need to take "ownership" of the various aspects of the program for which they are responsible. Included in this change is the concept that PMCD would no longer be directly responsible for the PLL program, but that a contractor (yet to be selected from the two operating contractors) would instead be responsible for it.

The philosophy and purpose of the PLL program are:

- to capture lessons learned during construction, equipment installation, systemization, operations, and closure, i.e., all phases of the operation
- to provide assistance to the sites and PMCD in assessing and utilizing these lessons and experiences
- to support PMCD's emphasis on safety and environmental compliance
- to reduce cost and schedule
- to provide information to decision makers.

The PLL program is a comprehensive, multicomponent activity that is distributed across all PMCD demilitarization sites and includes workshops, assessments, technical bulletins, directed actions and updates, programmatic planning documents, site document comparisons, critical document reviews, and a "quick react" feature (Box 4-1).

PLL PROGRAM DATABASE

The PLL program is a mechanism for developing and maintaining information associated with lessons related to preconstruction and construction activities, systemization, operations, and closure of the chemical demilitarization facilities. The majority of lessons learned are captured in the PLL database, which contains considerable information and is potentially an excellent resource for helping to maintain a high level of operational safety and security. However, so much information is present that plant personnel believe it is hard to identify what will be helpful in any given situation.

The information in the PLL computerized database is available to all participants in the PLL program. The database is searchable using both Boolean logic (and key word(s)) and a decision tree. Although other means of communication exist for discussing the operational and safety issues arising at the demilitarization facilities (described below), essentially all the information is contained in the database. The data are continuously updated and include information from workshops since 1994 and document reviews before that date. Not all PLL program components that lead to data included in the database were in place in 1994, and some have been improved since their inception. For example, workshops, critical document reviews, quick reacts, and the PLL oversight board were initiated in 1994; the technical bulletin, in 1995; and operational assessments, in 1996.

The issues database was first provided to the chemical demilitarization sites in 1997, the programmatic planning documents became available in 1997, the site documents comparison began in 1998, the directed action philosophy was revised in 1999, the engineering change proposal (ECP) review process (which began in 1987) was integrated with PLL in 1999, and the lessons learned database (a different way of sorting and accessing the information) was started in 2000-2001. The PLL team has also developed a help line to facilitate easier use of the information.

Data in the PLL database are accessible to the following staff:

1. PMCD home office, which includes stockpile disposal, alternative technologies, non-stockpile materiel, cooperative threat reduction, support offices, and contractors.
2. Project Manager Chemical Stockpile Disposal (PMCSD) and Project Manager for Alternative Technologies and Approaches (PMATA) sites, which include field offices and site systems contractors.
3. Other stakeholders, including operations support command, U.S. Army Corps of Engineers, U.S. Department of Health and Human Services, U.S. Army Center for Health Promotion and Preventive Medi-

BOX 4-1 Additional PLL Program Components

In addition to the computerized PLL database, the PLL program has several other significant components, including workshops, assessments, technical bulletins, directed actions and updates, programmatic planning documents, site document comparisons, critical document reviews, and a "quick react" feature. A brief description of these components follows.

- *Workshops* enable communication between and among PMCD personnel (including sites) and are the basis for information that ultimately is included in the PLL program database. Facilitated by a person knowledgeable about the issues (but not a decision maker) and usually from either PMCD or SAIC, the workshops are essentially focused technical meetings held in person or via teleconference or videoconference. Sample workshop topics include incinerators and secondary treatment support systems; general operations maintenance and training; personal protective systems; environmental, laboratory, and monitoring procedures; safety, surety and security; quality assurance/quality control (QA/QC); construction; systemization; public outreach; trial burns; and information management systems.
- *Assessments,* relatively detailed studies of an issue such as management or very technical topics, are intended as a means of rapidly starting an effort. A topic is developed by the government, and SAIC follows up in planning and execution with appropriate teaming partners (approved by the government).
- *Technical bulletins* are published quarterly and about 1000 copies are sent to various stakeholders. Each site makes additional copies as needed. The bulletins contain information that is not in the "quick react" category (discussed below) but requires attention before the next workshop, information that is of general interest but is not likely to be a workshop topic, or in some cases information that supplements the workshop discussions.
- *Directed actions and updates* transfer information or request that it be sent from or to the chemical demilitarization sites. Originating primarily in the workshops and/or quick reacts, directed actions and updates can also come from the PLL oversight board, critical document reviews, and other similar activities. The directed actions are assigned by PMCD managers and tracked by the PLL team until they are acted on. The responses are reviewed by the PLL team and incorporated into the PLL database.
- *Programmatic planning documents* that are maintained by the PLL team include a chemical demilitarization operations manual, PMCD management plan, PMCD information management plan, guide to systemization planning, guide to emergency response planning, and guide to closure planning (the last in draft form as of October 2001). The PLL team incorporates into these documents the results of lessons learned, adds new requirements (including applicable regulatory requirements), and comments on the cost/benefit considerations of producing a new version of any document.
- *Site documents comparison* involves PLL team review of new documents prepared at a site and comparison to previously approved documents from an earlier site and to programmatic guidance. The PLL team provides comments and recommendations, but implementing them is not mandatory at the site level.
- *Critical document review* is done by the PLL team for documents provided by the government, including reports of unusual occurrences or events, safety reviews (including near-miss advisories), reports of nonconformance or noncompliance, reports of test results, audit surveillance inspection reports, daily and weekly operating reports, and campaign reports. The purpose of these reviews is to identify lessons that will be added to the PLL database, update the database as needed, introduce appropriate topics for discussion at workshops and, if needed, recommend direct actions to secure further information.
- *Quick react* involves passing critical information—"changes to processes or equipment that affect operational safety, [or] environmental protection, or have the possibility to cause substantial equipment damage" quickly to the other affected parties. The site project manager or the chief of the operations division is responsible for and empowered to designate a lesson as critical. The time frame for a quick react is 24 hours. The quick react process consists of the following:

1. The site project manager (or the chief of the operations division) designates an issue as quick react.
2. The site faxes the information to PMCD and the PLL team (using a specific, designated form).
3. The site calls the chief of the operations division and the PLL contractor staff (who have 24-hour pagers).
4. The PLL team then conducts a data search and obtains any needed backup data, provides recommendations(s), faxes the government decision, confirms receipt, puts the data into the database, and tracks the directed actions.

The actual course of action is determined by PMCD operations management.

cine, U.S. Army Materiel Systems Analysis Activity, Edgewood Chemical and Biological Center, and regulators.

As of August 2001 the database was organized as an "issues" database and included about 3200 items from which users can choose to determine lessons applicable to their particular problem. Currently the PLL team (SAIC) is developing a new way to present the data and estimates that 5000 lessons will ultimately be available from the issues represented to date.

Although not specifically categorized as such, a significant number of laboratory issues are included in the database. Until recently JACADS provided most of the cases, but TOCDF is now providing most of the issues. The database also includes lessons from Anniston, Pine Bluff, Aberdeen, Newport, and Umatilla, all of which are currently under construction or undergoing systemization. The PLL team categorized these lessons as design, 341; systemization, 687; operations, 843; and closure, 241. Of these, 196 are categorized as maintenance and 202 as training lessons. Prior to 1999, the ECPs were handled in a separate manner, but all ECPs have now been captured in the database. Permitting issues are also included in the database.

When the PLL program began in 1994-1995, the major source of issues was the review of documents (event reports, end-of-campaign reports, inspection reports, and so on.). Now most of the information comes from the facilitated workshops run by the PLL program, which allow input and peer review by multiple program personnel with expertise in the subjects under discussion. The initiators of the information (subjects) are primarily the chemical demilitarization sites, but some issues come from other program participants. As currently operating, the decision process used to determine the ultimate content of the PLL database is as follows:

1. PMCD approves the list of topics (subjects) used at a facilitated workshop.
2. Twice a year the PLL team holds workshops for environmental and environmental oversight topics.
3. The minutes from the workshop are prepared, reviewed, and tentatively approved by SAIC. These minutes are sent to PMCD for its review and approval.
4. PMCD then makes the final decision before the minutes and lessons learned are entered into the database.
5. The database is distributed as a CD-ROM to each chemical demilitarization site to be loaded onto its local area network. It is not available on the Internet or on a wide-area network.

There is no mechanism to track the use of the data, but SAIC stated to the committee that use of the data is extensive at the engineering change proposal (ECP) level, as well as at the chemical demilitarization sites during start-up and operation. The committee found no accurate means for assessing this assertion other than the use of anecdotal information. When queried, some operators were unaware of the database and its uses.

SAIC is in the process of prioritizing the data so that the highest-priority issues will require a response from the site. At present, a site does not have to respond, since there are too many issues in the database relative to staffing levels at the site. Additionally each ECP approved by any site is discussed at a biweekly ECP review teleconference. At a subsequent teleconference the sites inform the PLL team of what action will be taken regarding the ECP. These appear to be among the few issues that are handled in this more structured manner. The ECP review process consists of the following steps:

1. The sites approve the ECPs and forward approved ECPs to the PLL team.
2. The PLL team researches related issues and ECPs (using the database) and sends the ECPs and accompanying information to the other sites.
3. The ECP review team, which includes representatives of the PMCD office, demilitarization sites, Army Corps of Engineers, and the PLL team, conducts biweekly teleconferences and puts the decision documentation into the database.

The PLL database and PLL concept reflect a systematic effort to take advantage of lessons learned in one chemical demilitarization facility and use the information at another facility. At present only one facility is operating (TOCDF), and one is undergoing closure (JACADS). As more facilities come on line it will be more difficult to track the data and ensure that the most important issues are addressed at all sites. PMCD will need to strengthen the communication and implementing mechanisms in the near future. PMCD (SAIC) is currently developing a set of criteria for prioritizing the information in the PLL database. The intent is to create a few categories of issues (lessons), sorted according to relative importance. For instance, those with the highest priority (for example an important, operational safety directive) should probably be available and implemented at all facilities.

RESULTANT CHANGES

At JACADS and TOCDF, operations personnel did not appear to generalize lessons learned beyond the immediate equipment and task in the original incidents. There is room for making much wider use of these valuable lessons, such as by "mining" the information in the PLL database to detect patterns that may underlie several incidents. The effort to prioritize the data is a good start toward increasing the information's usefulness. PMCD could also make better use of information available from industries such as the chemical and petroleum manufacturing sectors. Both have very active trade associations and routinely share information regarding safety procedures and good operating and maintenance practices among different companies.

The destruction of chemical weapons was first begun at JACADS and its design was based on equipment and procedures developed at the Chemical Agent Munitions Disposal System (CAMDS) at Deseret Chemical Depot, Utah. Many design changes were made after operations had begun, some in response to chemical events, but most to correct recognized problems with the original design. (Both types are included in the PLL database.) For all operating chemical facilities, design changes are part of a continuing process aimed at taking advantage of lessons learned from ongoing operations, new technology as it is developed, or better procedures developed at a plant or transferred from another facility.

Many design changes have also been made to improve productivity (e.g., inclusion of the hot slag withdrawal system on the liquid incinerator (LIC) secondary burner, and the process, currently under review, for freezing the M2A1 projectiles at Anniston before disassembly to minimize spilling, and subsequent cleanup, of mustard agent). Design changes to improve operating safety, however, are not as readily identified except in direct response to a chemical event. For instance, the airflow systems handling ventilation throughout a plant as well as combustion air will have variable-speed motors driving the fans, allowing improved control of airflow, particularly at low rates (combustion airflow control was a problem for the operator during the May 8-9, 2000, incident at TOCDF).

A large number of changes have been made to operating procedures and equipment in response to the PLL program and based on incident reports from JACADS and TOCDF. Of the 24 recommendations for change resulting from the May 8, 2000, event at TOCDF, for example, all have been examined, although not all have required action at the newer plants because of differences in the feed mix and in the plant designs. In the committee's view, some of the more significant changes made in response to the PLL are as follows:

- Staggered automatic continuous air monitoring system (ACAMS) monitors are now being installed in exhaust ducts, to shorten the time for detection of any release of agent.
- The deactivation furnace system (DFS) cyclone is contained in an enclosure that is monitored by an ACAMS.
- There is a carbon filter on the incinerators' exhaust.
- As a result of the JACADS waste-bin event (discussed in detail in Chapters 2 and 3), drip trays have been added to rocket and mine lines, a search is on for combustible spill pillows, and spill pillows will generally be treated in the metal parts furnace, not the DFS.
- The large isolation valves on the individual heating, ventilation, and air conditioning (HVAC) carbon filter banks now have a small "bleed" valve, connecting to the exhaust flow, to maintain the filter bank at negative pressure even if it is temporarily out of service and to prevent migration of agent from the filter bank to the connecting vestibule. (Such migration of agent has been a problem in the past.)

Chapter 1 discusses the systems hazards analysis (SHA) performed for TOCDF. A primary purpose of a standard hazardous operation (HAZOP) analysis is to learn to anticipate where safety may be compromised. There have been many changes to the original design (see footnote 1, Chapter 5), some identified above and all included in the PLL database. It is not apparent that each of these design changes has been subjected to the appropriate level of HAZOP analysis. In view of the challenging nature of the chemical weapons disposal program and its perceived potential for harm, this aspect of the design process needs particular and ongoing attention.

It is common practice in industry for people who do the design and initial HAZOP analysis to be included on the plant start-up team. The people who did the actual detailed design work and participated in the HAZOP studies done as a part of the design process should also play a strong role in operator training in the use(s) of the HAZOP procedures and information. It is also common industry practice for companies to share nonproprietary information about safety issues, operating procedures, HAZOP findings, and so on. PMCD could make better use of the experiences of other industries, such as the chemical and petroleum refining industries, in the benchmarking of its procedures and processes.

5

Preparing for Potential Future Chemical Events at Baseline Chemical Demilitarization Facilities

SUMMARY OF CHEMICAL EVENTS ANALYSES

The committee's analyses of past chemical events at Johnston Atoll Chemical Agent Disposal System (JACADS) and Tooele Chemical Disposal Facility (TOCDF) indicate that the causal factors are similar to those associated with breakdowns of other safety-critical systems. Release of chemical agent may be triggered by equipment design flaws and failures, by procedural deficiencies, and by human actions—i.e., by both latent and active failures (see Chapter 2).

The task of dismantling and destroying chemical weapons is inherently hazardous, but the Program Manager for Chemical Demilitarization (PMCD) has incorporated extraordinary safety precautions into both plant design and personnel training. The chemical demilitarization incineration plants are virtual fortresses built to withstand the consequences of accidents, and, to date, releases of chemical agent from these facilities have been rare, isolated events involving only small amounts of agent, even under upset conditions (NRC, 1996, 1997, 1999a). State-of-the-art quantitative risk assessments have determined that the major hazard to the surrounding communities arises from potential releases of agent from stockpile storage areas, not the demilitarization facilities (U.S. Army, 1996a; NRC, 1997; see also Chapter 1 and Appendix E). Further, to date by far the largest releases of agent have occurred in the storage areas, as described in Chapter 1.

The Army has sought to build in the process of learning by experience to avoid accidents where possible, and to avoid repeating them in any case. The centerpiece of this effort, the programmatic lessons learned (PLL) database, is admirable as a personnel-training tool but requires further modification to improve its accessibility (see Chapter 4). Despite considerable effort in plant design and personnel training, mistakes have been made and problems have occurred in the chemical demilitarization process.

The Army has established extraordinarily low agent threshold concentrations to trigger site alarms and a subsequent shutdown of the plant (see Chapter 1). While laudable as an effort to protect worker and public health, these overly sensitive alarms introduce their own kinds of operating problems. Difficulty in reliably detecting agent at such low concentrations leads to recurring false positive alarms. It also means that alarms triggered by chemical events in which agent levels stay near threshold will actually pose no risk to the worker or the public.

Given the inherent complexity of the chemical demilitarization task at the assembled weapons stockpile sites, it is almost certain that new problems will continue to arise, particularly from aging and deteriorating weapons and the challenges of demilitarization plant closure and decommissioning. There will be future chemical events, and serious consequences to both plant personnel and surrounding communities cannot be ruled out. This chapter focuses on prudent ways to reduce their number and to minimize their consequences.

CHEMICAL EVENT RESPONSE AND REVIEW BY MANAGEMENT

Army Regulation 50-6 presents in detail the response to a chemical event and its reporting expected from the depot commander (U.S. Army, 1995). The objective is to:

> . . . encompass those actions to save life, preserve health and safety, secure chemical agent, protect property, prevent further damage to and remediate the environment, and help maintain public confidence in the ability of the Army to respond to a military chemical accident or incident. . . . The major army commands (MACOM) commanders will establish procedures to review each chemical event and to initiate safety investigations when warranted . . . .

The extent of the review process generally varies with the seriousness of the incident. The review process for a serious incident can be quite lengthy. Every chemical event should be investigated promptly, particularly those considered potentially or actually serious. Memories of the event will change with time. Having people identified in advance as potential candidates for a review team would appear worthwhile.

One of the objectives of Army Regulation 50-6, stated above, is to "help maintain public confidence." The committee believes that building trust requires regular and reliable communication between the Army and the communities around the demilitarization plants. It does not appear that these communities feel that such communication has been achieved. Public trust is not easily established and is very difficult to rebuild once lost. The recent report of the U.S. Commission on National Security (a commission headed by former senators Gary Hart and Warren Rudman) comments on the general lack of confidence in federal employees (USCNS, 2001). This general lack of confidence, exacerbated by the unfortunate pattern of interactions between PMCD and external stakeholder groups (NRC, 1996), has created a serious deficit of trust in the Army's chemical demilitarization program on the part of important segments of the public. In addition to addressing the public's lack of confidence in federal officials, at some sites PMCD must also deal with public distrust of state and local officials. A recent NRC letter report (NRC, 2000c) points out that:

> . . . open, two way communications between PMCD and stakeholders are necessary, but insufficient. PMCD needs to encourage public trust in official representatives of the public (i.e., Citizens Advisory Commissions and local regulatory bodies) as much or more than it needs to build trust in the Army.

The memorandum of understanding between TOCDF and Tooele County (see Appendix G) should help build confidence that public officials are fully informed and responsive to chemical events, thereby contributing to building trust. This approach might serve as a model for other communities with similar concerns (Utah DEQ, 2000b).

BUILDING ON THE RESULTS OF RISK ASSESSMENT

Risks associated with the chemical demilitarization facilities have been studied in depth, through quantitative and health risk assessments and systems hazards analyses (see Chapter 1). The quantitative risk assessment, in particular, is a living document, subject to change as new information arises or facilities or operations are altered. It provides excellent guidance on where risk is the highest, and thus where the greatest care is needed. The Army's "Guide to Risk Management Policy and Activities" provides a process for managing risks, particularly when changes are made, and for communicating information on change to the public (U.S. Army, 1997b).

Understanding and building on the results of risk assessment implies more than knowing the summary numerical results of quantitative and health risk assessments. It also requires knowing the details, including the assumptions, simplifications, and omissions, of the analyses. The results must be viewed in the full context of the risk assessment, as well as in the context of the actual safety performance of the plant. This perspective must be accompanied by a better understanding of explicit and implicit uncertainties.

Understanding the results of risk assessment also means knowing the significant contributors to risk, i.e., knowing how improved performance can reduce risk and how degraded performance could increase risk. With this knowledge:

- Managers and workers can develop options for reducing risk or for ensuring that risk does not increase. They can also consider how proposals for change affect risk.
- Workers, emergency response personnel, and others can better understand their personal risks and how best to protect themselves and each other.
- Emergency preparedness managers can focus their planning and training programs on the most important scenarios or sources of risk to the surrounding communities.
- State and local officials can provide more informed oversight in their decision making.
- Everyone can participate knowledgeably in the risk management process.

Quantitative and health risk assessments are complex and, of necessity, include simplifications. The plant safety professionals should review the assessments thoroughly to be aware of their basic assumptions and/or limitations. Plant operating requirements may change, and changes need to be viewed in the light of the risk assessments.

Several lessons can be learned about risk management from thinking about possible responses to certain kinds of chemical events.

- *False positive alarms.* The history of false positives has contributed to a number of chemical events as described in Chapter 2. These result from a mindset that develops in operators. Faced with a series of false positive alarms, they tend to disbelieve future alarms, at least to the extent that they seek confirmation before taking action. A question that has been raised is whether a similar complacency could develop among emergency response managers and even the general public? If they too are subjected to false alarms, they may delay ordering or responding to orders for evacuation or sheltering. Generally, these people have not been subjected to the false alarms, but if it should happen, similar problems could arise.

- *Evacuation versus sheltering.* At some sites, there has been controversy over the question of evacuation versus sheltering. Countering the belief that evacuation is always the safe path are at least two circumstances. First, evacuation itself can create hazards. It disrupts the economy and daily life and can create high stress. It has led to injuries due to traffic accidents and improper use of safety equipment. Second, analyses by Chemical Stockpile Emergency Preparedness Program (CSEPP) planners have shown that for some release scenarios, evacuation can place the evacuees directly in the path of the hazardous plume (Blewett et al., 1996). For some scenarios, sheltering in place (remaining indoors with the doors and windows sealed) as the plume passes, followed by evacuation, can greatly reduce exposure. Continued sheltering after the cloud has passed may lead to exposure as severe as being caught in the plume. In these cases, sheltering as the cloud passes, followed by evacuation through contaminated areas, can be the most effective protective action.

BUILDING A SAFETY CULTURE

TOCDF has clearly made an effort to promote plant safety. Two examples are (1) the use of the Safety Training Observation Program (purchased from the DuPont Company) and (2) the use of the Voluntary Protection Program developed by the Occupational Safety and Health Administration (OSHA). A good safety organization on paper, however, does not ensure a high-quality safety culture. Some of the past events at both JACADS and TOCDF arose from obviously poor safety practices. The recordable injury rate (RIR) at TOCDF, for example, has been unimpressive (Chapter 4). The NRC has emphasized the need to focus on safety with constant attention to detail, starting with a complete and persistent commitment from management (NRC, 1999a).

OPERATIONAL CHANGES

It is clear that (1) serious mistakes have been made in chemical demilitarization plant operations in the past and (2) strict standards of operating practice have not been uniformly enforced (see Chapters 2 and 3). These are failures of management to fulfill their responsibilities. Improvement will come only with serious management effort, significantly greater than in the past. Strong safety cultures and an adherence to defined operating procedures have been established in other industries. The goal for chemical demilitarization plants should be to match the best achieved in industry.

A criticism that is easily voiced but difficult to respond to is the general acceptance of the status quo by chemical demilitarization operating people and management. Changes are made in response to chemical events or obvious operating difficulties, but based on the committee's site reviews, a culture of questioning processes and constantly improving operations does not seem to exist. To be fair, it is clear that plant management is aware of the importance of being proactive on safety, rather than being reactive only. Certainly there has been real improvement in plant layout, equipment, and so on (see Chapter 4). Based on the committee's observations and discussions with operating personnel, TOCDF is clearly a better designed and engineered plant than JACADS, and the third-generation incineration plants, as exemplified by the Anniston Chemical Agent Disposal Facility, appear to be a significant improvement[1] on TOCDF. Many of these improvements were made by seeking better ways of doing things, and anticipating possible future problems rather than reacting after a problem has occurred. The committee encourages a continued vigorous questioning of plant operation and equipment by management and operating personnel. This open-minded, questioning approach should apply to operating practices and even equipment design.

---

[1]Although the basic processes for weapon destruction will remain the same three lines of incineration as at TOCDF and JACADS (a furnace for injecting and burning liquid agent, a rotary furnace for propellant and explosive materials, and a furnace with a moving conveyor primarily for metal parts), improvements have been made compared with TOCDF and JACADS. For example:

- The pollution control systems of the new plants will include activated carbon filters for the incinerator exhaust gas. This is fairly new technology, not in common use when JACADS and TOCDF were designed. Trial burn data on those two early plants showed that carbon filters were not needed to meet environmental standards. More recently, however, some samples of mustard have shown unexpectedly high levels of mercury that could be a problem in exhaust emissions. Carbon filters represent the technology of choice for handling this problem. Other changes in the pollution abatement system are required to accommodate the carbon filters. The exhaust gas must be cooled and its humidity reduced to maintain the carbon filter's function.
- The ventilation air through the plant as well as the combustion air will have variable-speed motors driving the fans. This should be a great improvement in controlling airflow rates, particularly at low rates (a problem in the May 8, 2000, TOCDF incident). The technology for doing this with very large motors was just being introduced when TOCDF was designed and was not included.
- Isolation valves are included in the duct between the DFS burner and after burner. (The same valve was added to TOCDF after the May 8, 2000, event.) They should permit improved control during start-up.
- The DFS tipping gate has been redesigned to prevent jamming (part of the problem in the May 8, 2000, event).
- The large isolation valves on the individual HVAC carbon filter banks have a small "bleed" valve connecting to the exhaust flow in the new plants. The purpose is to maintain the filter bank at negative pressure even when the filter is temporarily out of service. This should prevent migration of agent from the filter bank into the connecting vestibule when the filter is out of service, a problem in the past.
- The DFS cyclone is in an enclosure that is to be monitored with an ACAMS and that has a carbon-filter on an exhaust. (This modification was made in response to a JACADS event where VX was detected on the cyclone ash.)

The 2000 letter report of the NRC Stockpile Committee recommended a similar open-minded approach for a public involvement program (NRC, 2000b), naming as one requirement for a successful program "the capability to identify (even anticipate) serious problems and the flexibility and creativity to address them." The current study suggests that this approach is also needed in plant operations and technology.

A better understanding of the limitations of plant equipment might also be helpful. In the May 8-9, 2000, incident at Tooele, for example, there were some serious operating errors. But they were compounded by the operator's struggle to bring the system back under control. It is the judgment of the committee that some technical education and more hands-on testing on the system simulator would have helped (see Chapter 2).

There are usually surprises when new processes are first tried. In view of the particular sensitivity of the chemical agent disposal program, the committee emphasizes the need for a hazardous operations (HAZOP) analysis for any new process (see Chapter 4). A HAZOP analysis by suitably trained people, and with input from operating people, could be particularly useful: it might identify problems and at the same time warn the operating people about what to expect.

New plant start-up represents a special problem with inexperienced people. Trial burns with surrogate feeds and with the pollution abatement system in full operation, as well as disassembly trials with blank munitions, should provide substantial operating experience before any chemical agent is fed to the process. It is fairly common experience in industry to include design people on start-up teams for new facilities. As suggested earlier, their detailed knowledge of the process equipment and its limitations could be helpful to the operating people.

WORKER EDUCATION, TRAINING, AND INVOLVEMENT

Safe plant operation depends on an educated, well-trained staff. The risk to workers in an incineration plant is greater than the risk to the public (NRC, 2000c). Training should emphasize that processing agent demands a mind-set that always accepts a positive analysis as "real" until proven otherwise.

One approach to safe operation is through the use of standard operating procedures (SOPs). These have been used extensively at JACADS and TOCDF. The most serious chemical events of the past have occurred, however, when there was no SOP. There will always be combinations of circumstances for which no SOP has been written and the operating people must rely on knowledge-based decision making. Even with SOPs, there is no guarantee that mistakes will not occur. It is vital that decisions be made on the basis of accurate operational knowledge. Operating people should know their equipment and its limitations. They need to know the why of their job as well as the what. Bringing the systems engineers with design knowledge into the training program could help convey that knowledge to the operators. These engineers are probably in the best position to know the equipment and its characteristics and limitations, information that plant operators need when unusual or unexpected conditions occur. Many plant operators seemed to the committee to have only a superficial knowledge of the operating principles and data processing algorithms of important process instrumentation and controls. But such knowledge is crucial to determine how to interpret reported instrument console readings during upset conditions which may exceed the normal ranges over which key instruments are calibrated or can be expected to operate reliably. A careful walk-through of any new procedure should precede its start-up.

The Army's more recent quantitative risk assessments (QRAs) include detailed human reliability analyses that identify potential human performance problems. Bringing this information into the training program will provide operators with a view of what activities are especially vulnerable and why that is so. In addition, training simulators, which mimic the operation of the various components of the instrument and control systems and demonstrate the effects of various operator actions or inactions, are now being provided in the chemical demilitarization plants. Targeted training with simulators and knowledge-based thinking exercises on plant operation need to be developed.

Training on overall plant operations should cover everyone in the plant and analytical laboratory, not just the operating and control people. However, this training has to be tailored to the specific jobs and knowledge levels of each group of workers. Workers need to understand how what they do fits into the overall operation and how things going wrong in their operations affect the whole plant and the likelihood of accidents and releases. The QRA and HAZOP analysis are a good potential source of this information.

Some of the reports of operational mistakes coming from within the plant and circulated widely within the affected communities have come from people who are simply uninformed and do not know normal procedures. Box 5-1 provides examples of such uninformed observations. Chemical demilitarization plants are complex. A better knowledge of the complexity of the plant and the care and design that have been incorporated may instill pride in being part of the important national effort of weapons disposal. The potential costs (e.g., lost trust) of having the local public alarmed by reported misperceptions of uninformed workers can be substantial.

DESIRED PRINCIPAL-*AGENT* INTERACTIONS

It is imperative that officials at the chemical demilitarization facilities communicate openly, frequently, and in a timely fashion with nearby residents and officials. The pattern of communication with and responsiveness to the local

> BOX 5-1 Examples of Observations That the Committee
> Concluded Were Uninformed
>
> *"December 9, 2000—Agent break through in HVAC filter bank. ACAMS readings of 3.01 [TWA]."*
>
> *"October/November 1997: Sources inside TOCDF (who wish to remain anonymous) communicated to CWWG [Chemical Weapons Working Group] several shutdowns/incidents at TOCDF due to computer malfunctions, slag build-up in the PAS, numerous agent migrations within the facility, and alarm ring-offs in the common stack, MDB [munitions demilitarization building] and HVAC stack (averaging 2-3 per week)."*
>
> These entries suggest that agent may have been released through the heating, ventilation, and air conditioning (HVAC) filter to the environment. In fact, the HVAC was operating as designed. The carbon filter bank consists of six carbon beds, with exhaust gas flowing through all six in series. The gas spaces between beds 1 and 2, 2 and 3, and 3 and 4 are monitored by an automatic continuous air monitoring system (ACAMS) (on a timer). Eventually agent will break through bed 1 as that bed approaches saturation, and this is undoubtedly the "agent break through" referred to by the whistle-blower. Agent breakthrough of bed 1 usually follows many weeks of operation, and with the gas having to traverse 5 more beds the agent breakthrough of the first bed does not call for immediate shutdown. However, it does indicate that the carbon should be replaced soon.
>
> *A video given to the committee and referred to in Appendix C showed rockets being sheared and the pieces dropping to the deactivation furnace system (DFS) below.*
>
> *Approximately every $1^1/_2$ minutes a large cloud of condensing vapor, referred to by the citizen group as "agent volatization," rose into the picture, undoubtedly coinciding with opening of the gate to the DFS.* In fact it was a cloud of condensing steam, as cooling water from the shear blade and the sliding gate dropped into the hot furnace to be instantly vaporized.
>
> *"Site-masking alarm and/or stack alarm. Potential case of chemical warfare agent release or release of other related toxic chemicals (unidentified to date)."* [the most common incident listed by the CWWG (Appendix C)]
>
> It is almost certain that the ACAMS alarm was not due to agent, because there was no depot area air monitoring system (DAAMS) confirmation. The committee concluded that the event reports as written are misleading but considers them to be from a source unfamiliar with the stringent laboratory procedures used to analyze DAAMS samples taken coincident with each ACAMS alarm to confirm or deny the presence of agent and to attempt to identify the cause of the alarm in the absence of agent.
>
> NOTE: Observations quoted are drawn from the Chemical Weapons Working Group list of events provided to the committee (Appendix C).

public and local officials can have substantial effects. Beyond addressing the immediate health and environmental concerns posed by a chemical event, frequent and open dialogue can alter perceptions of risk and trust, influence demands for policy change, and mitigate undesirable effects on local economic growth and property values. As discussed in Chapter 3, the *agents* in the demilitarization process (regulatory agency officials, the Army, and contractors at the chemical demilitarization facilities) must gain and retain the trust of the principals (local public and the officials who represent them) in order to effectively destroy the chemical weapons stockpile in a safe and timely manner.

Absent complete trust, the mechanisms by which principals gain confidence in adequate performance by *agents* include effective monitoring of *agent* behavior and appropriate inducements and sanctions to obtain desired performance. The lower the level of trust, the greater the need for monitoring and incentives. At the same time, more stringent monitoring and incentives can limit the discretion necessary for *agents* to effectively and efficiently accomplish their complex task. The trade-off between effective monitoring and controls by principals over *agents* and optimal conditions under which *agents* can carry out the demilitarization task (where some discretion may be essential) requires engendering and maintaining a degree of trust by principals for *agents*.

Effective handling of the principal-*agent* relationship in the chemical demilitarization program setting appears to the committee to require (1) demonstrable and timely assessments of the problems leading to chemical events and means for their correction, (2) complete and timely disclosure of events by the *agents*, and (3) overview processes that assure principals of effective oversight.

In its assessment of chemical events (Chapter 2), the committee found specific instances (e.g., the TOCDF May 8-9, 2000, incident) that resulted in a damaging erosion of the confidence of principals in the monitoring and control processes. It is essential that plant operators remain cognizant of the needs of principals for high degrees of confidence in the monitoring and control protocols (incentives and sanctions) and mechanisms over the entire chemical demilitarization program. Apparent weaknesses or failures at one facility or in one phase of operations will be seen to carry over to others. Protocols for reporting and responding to events should stress meeting the needs of the array of external principals for assurance of timely, accurate reports of events and rapid, thorough assessment and corrections.

One important step to increase confidence in the monitoring process will be to ensure that representatives of principals (e.g., local stakeholder groups) are included in the teams assembled to investigate any serious chemical events. In addition, each site should develop clear and specific protocols that reflect the need to quickly, openly, and thoroughly inform all relevant principals of chemical events.

More broadly, program officials should consider ways in which principals and their representatives can participate in ongoing oversight efforts. The NRC has suggested elsewhere (NRC, 1999b) that representatives of the local public serve on monitoring teams whose purpose is to ensure that chemical weapons destruction processes (and the associated organizations) are operating as they should. Such an effort—ranging from temporary appointment of community observers on investigation teams to more permanent participation of community representatives in incident review boards—may increase the confidence of local principals that effective oversight is in place.

RAPID AND SAFE RESTART REQUIREMENTS

Restarts After Changeovers and Maintenance

The chemical weapons plants have very frequent shutdowns and restarts—"frequent" compared with industrial plants of comparable size. These shutdowns are required by the variable nature of the plant feed: a variety of weapon types with differing disassembly requirements, containing three different chemical agents. The times required for changeover have been estimated to be surprisingly long (U.S. Army, 2000c). For example:

- A change in agent:
  —17 weeks—the time required for decontamination, monitor conversion and baselining, and some equipment changeovers.
- A change in munitions type:
  —5 weeks without complete equipment removal (e.g., projectile to projectile types).
  —8 weeks with equipment removal (e.g., mines or rockets to projectiles).

There are other normal maintenance items that require extended shutdown periods but can probably be scheduled during other changeovers. For example:

- Slag removal from the liquid incinerator (LIC) secondary burner. [The slag removal system at TOCDF failed before the refractory failed, so that slag had to be removed manually.] Time required: 10 days. The experience at JACADS and TOCDF permits an estimate of the required frequency of slag removal, e.g., for TOCDF, after 250,000 lb of agent destroyed.
- Mist eliminator candle replacement (plugged during deactivation furnace system (DFS) rocket runs, probably due to fiberglass). Time minimized by having a spare eliminator on hand.
- LIC rebricking, maintenance that can also probably be planned ahead and done during "contingency time" (i.e., when the plant will be shut down for other activities such as agent changeover). Rebricking is needed after approximately 2,000,000 lb of agent (with decontamination fluid) have been processed.

A further complication arises from the age of the weapons as well as their varied history—"leakers" and "gelled agents" require changes in "standard" operations, for example. The shutdowns and restarts resulting from these feed stock variations can be planned for and shutdown times can be minimized.

The Operations Schedule Task Force 2000 recommended study teams to suggest how to minimize downtime (U.S. Army, 2000c); these teams should be very helpful. The committee suggests that industrial experience with carefully planned shutdowns for maintenance at regular intervals might be applicable. It is not clear that "project management," which has developed into a distinct engineering subdiscipline, is being fully integrated into the chemical demilitarization program. The suggested study teams noted above would represent a step in that direction.

Restarts After a Chemical Event

Major chemical events can impose further shutdowns with unpredictable shutdown times. Some of these have led to major structural changes and changes in some operating procedures. These changes stem from the incident reviews, and they all require regulatory approval. Shutdown times may be long, e.g., 4 ½ months for the May 8-9, 2000, incident at TOCDF. The Operations Schedule Task Force 2000 suggested that a 2-week outage every 6 months be included in advance planning, to accommodate unplanned major maintenance (U.S. Army, 2000c). The committee believes that this unplanned shutdown allowance is less than past experience would indicate is necessary, but these unplanned shutdowns should decrease with time, as operating experience is gained. There also may be opportunities for reducing the required shutdown times after such incidents. Maintaining a larger inventory of critical spare parts has been suggested as one strategy to reduce lost operational time.

Obtaining regulatory approval to restart after a chemical incident may cause delays, although the committee heard no specific complaints of this. The Army Audit Agency, however, has been critical of the chemical demilitarization program for its handling of funds, based in part on regulatory delays. Funds obtained for current planned programs could not be spent because of delays in regulatory approvals (U.S. Army, 2001g). It is not clear, however, that regulatory delay has been a serious problem in connection with unpredicted shutdowns, where there was no opportunity for advance planning.

Finally, as noted in Chapter 3, effort spent on the multiple investigations of the May 8-9, 2001, Tooele chemical event probably extended the post-event shutdown associated with that event unnecessarily. Preagreement at each demilitarization site on the composition of a joint event investigation team, representing all regulatory and operational stakeholders and chartered to produce a single, comprehensive investigation report, could save significant shutdown time and clearly focus all parties on the steps necessary to achieve safe restart of operations after future chemical events.

6

Findings and Recommendations

Based on its review and analysis in Chapters 1 through 5 of incidents at two operating chemical demilitarization sites, JACADS and TOCDF, the Committee on the Evaluation of Chemical Events at Army Chemical Agent Disposal Facilities developed the following findings and recommendations.

Finding 1. Despite considerable Army security and stewardship activities, the remaining chemical weapons stockpiles are significant hazards to the communities surrounding them. The potential for significant release of agent to the atmosphere, triggered by either accidental or deliberate detonation of agent-loaded munitions within storage igloos, constitutes the greatest risk to the public. Accidental or deliberate release from a chemical demilitarization facility, while potentially serious, is a lesser threat because the agent inside the facility is maintained under stringent and effective engineering controls and because there is substantially less agent present in the demilitarization facility at any given time than there is in the storage facility. While chemical demilitarization operations at both Johnston Atoll Chemical Agent Disposal System and Tooele Chemical Disposal Facility have released small amounts of chemical agent into the environment, these releases were negligible compared with environmental releases from chemical weapons stockpiles (U.S. Army, 2001e). The rate of agent leaks and releases from storage facilities does decrease significantly as the stockpile is processed.

Recommendation 1. The destruction of aging chemical munitions should proceed as quickly as possible, consistent with operational activities designed to protect the health and safety of the workforce, the public, and the environment.

Finding 2. The criteria used by the Army to identify and determine the severity of the impact of an event are site and time specific, and the event classification decision is made at the discretion of the Depot Commander. The recognition of a chemical event is often subjective, and the tendency of personnel to discount initial indicators because of frequent false positive automatic continuous air monitoring system alarms is a persistent problem in declaring a chemical event. The lack of uniform criteria can result in inconsistencies between and among sites that make it difficult to compare and analyze events and that constrain and discourage the application of lessons learned to other locations and situations.

Recommendation 2. The Army should establish a consistent set of criteria to be used by all chemical-agent-processing facilities to ensure uniformity in the classification of events, and to facilitate event analysis and comparison.

Finding 3. Risk assessments, including the quantitative risk assessment and the health risk assessment, are critical inputs to the dialogue necessary to ensure adequate public involvement in, and understanding of, chemical demilitarization activities. A prudent balance between the public's right to know the risks they face and the need to protect sensitive information will be an ongoing challenge for the chemical demilitarization program. Without adequate risk information available to the public, it will be difficult to develop or maintain the level of public trust necessary for the Program Manager for Chemical Demilitarization to accomplish its mission.

Recommendation 3. The Army should continue its practice of making available to the public the results of its quantitative risk assessments and health risk assessments for each chemical demilitarization site.

Finding 4. Of the wide range of Program Manager for Chemical Demilitarization risk analyses, the quantitative risk assessments (QRAs) are most closely linked with chemical events. They calculate the frequency and consequences of modeled events, and their analysts study real operational

51

events to help ensure the completeness of the models. The QRAs, and an understanding of their results, provide a framework for managing the risk from chemical events. As concern has focused on worker risk as well as risk to the public, recent QRAs have added extensive human reliability analysis—analysis of the human actions that can lead to a chemical event. While hazard analysis is a qualitative analysis primarily of single-failure events, it provides a catalog of possible failures that QRA analysts can use to improve the completeness of their models. Hypothesized chemical events, including those that might arise from sabotage, terrorism, and war, can also be incorporated into the QRA scenarios to determine their range of consequences. Actual events can be used to test and improve the completeness of the QRA and continue the effort to improve the human reliability analysis and focus on causal factors.

Recommendation 4. The quantitative risk assessment (QRA) for each chemical demilitarization site should be iterative. Actual chemical events should be used routinely to test the completeness of the QRA, which should be routinely utilized to hypothesize the frequency and consequences of chemical events. The Program Manager for Chemical Demilitarization and the U.S. Army Soldier and Biological Chemical Command should use the QRAs to evaluate measures to control future chemical events. The Army should also consider using QRAs to examine scenarios associated with sabotage, terrorism, and war.

Finding 5. Alarm thresholds for airborne agent monitoring used in the Army's chemical demilitarization program are very conservative (i.e., 20 percent of the applicable control limit, resulting in alarm thresholds that, depending on the agent, are either below or only moderately above the level of agent deemed safe for continuous exposure of an unmasked worker over an 8-hour shift.) These alarm thresholds are near the detection limits for the automatic continuous air monitoring system (ACAMS). As a result, there are frequent false positive alarms as well as actual alarms for events that pose no threat to workers or the public (NRC, 2001a). These conservative stack-monitoring thresholds ensure that no significant amounts of agent can be released into the ambient air without the facility alarming and the agent incineration feed automatically terminating. In-plant air breathed by unmasked workers and the output of the scrubbing system for air exiting the chemical demilitarization plant are monitored at similarly conservative thresholds.

Recommendation 5. The Army should maintain conservative chemical demilitarization exhaust stack and in-plant airborne agent exposure thresholds. If current limits for exposure to stockpiled chemical agents are further reduced, the Army should not further reduce existing monitoring thresholds unless chemical agent monitors can be made both more sensitive and more specific so that lower thresholds can be instituted without significant increases in false positive alarm rates or unless health risk assessments demonstrate that lower thresholds are necessary to protect workers or the public.

Finding 6. Relatively frequent false positive ACAMS alarms for airborne agent and the lack of true real-time (<10 s) monitoring for airborne agent have long been a concern of National Research Council (NRC) committees assessing and examining the chemical demilitarization program (NRC, 1994, 1999a, 2001a). Improvements in the sensitivity, specificity, and time response of the ACAMS system and the development of an additional airborne-agent-monitoring technology robust at the parts-per-trillion level have previously been recommended. (Recent NRC reports have also noted the poor state of agent-monitoring technology for liquid waste streams and solid materials suspected of possible agent contamination (NRC, 2000a, 2001a).) Although the Program Manager for Chemical Demilitarization has made some efforts to develop better agent-monitoring technology, results to date have been disappointing. Development and deployment of airborne-agent monitors with shorter response times and lower false alarm rates would enhance safety and reduce the tendency to discount agent alarms.

Recommendation 6. To reduce the rate of false positive alarms for both airborne and condensed-materials agent contamination, the Program Manager for Chemical Demilitarization and the relevant Department of Defense research and development agencies, such as the Army Research Office, the Army Research Laboratory, the Defense Advanced Research Projects Agency, and the Defense Threat Reduction Agency, should invigorate and coordinate efforts to develop chemical agent monitors with improved sensitivity, specificity, and time response. These efforts should be coordinated with, and take advantage of, the increased level of interest in and increased resources available for developing chemical weapons detectors for homeland defense.

Finding 7. Chemical demilitarization facility and process design has contributed to the mitigation of incident severity in that, for most of the incidents examined by the committee, engineering controls functioned as designed. These incidents have been investigated honestly and thoroughly using straightforward techniques, but investigation could benefit from the use of other methodologies such as event tree analysis and human factors engineering to aid in understanding the complete set of causal factors associated with each incident.

Recommendation 7. Incident investigation teams should use modern methodologies of incident investigation routinely at all chemical demilitarization sites to help uncover a broader set of causal and contributing factors, and to enable greater understanding of the interrelationships between and among these factors. Experts in human performance modeling should be included on any incident investigation team. A standing incident review board at each site should be established to identify chemical events requiring in-depth investigation and to ensure that the lessons learned appropriately influence ongoing operations. These boards would meet regularly to review accidents and incidents, including chemical events, and would be fully informed of any findings and recommendations made by chemical event investigation teams.

Finding 8a. Repeating patterns of causal factors evident in the incidents at Johnston Atoll Chemical Agent Disposal System and Tooele Chemical Disposal Facility reviewed by the committee included, in particular, deficiencies in standard operating procedures, design failures, and understandable, although inappropriate, assumptions made by operations personnel. In part, these inappropriate assumptions grew from development of dangerous mind-sets associated with frequent false-positive alarms. Repeating patterns of causal factors in most incidents do not appear to have been used by management to generalize incident findings beyond the immediate context of each incident.

Finding 8b. The programmatic lessons learned (PLL) database is a large undertaking, and the Program Manager for Chemical Demilitarization is to be commended for creating it. However, if the data were organized in a different manner that included a priority system and the operators were aware of its uses, the database would be more useful. "Mining" of data might yield patterns in events and information that would allow broader generalization and understanding of causes derived from specific information on individual incidents. To this end, experts in each area of use have to specify exactly what they need to find in the data, before programmers develop software to search and set priorities.

Recommendation 8a. The Program Manager for Chemical Demilitarization should analyze all chemical-agent-related incidents at chemical demilitarization plants for patterns of causal factors and should institute program-wide actions to address the causes found.

Recommendation 8b. Any improvements made in investigation procedures should become part of a systematically organized programmatic lessons learned (PLL) database that makes information easier for the non-expert to find and/or use. This can include prioritization and developing a drop-down "tree" list. Lastly, the Program Manager for Chemical Demilitarization should ensure that, at the plant level, the data are available to, known by, and useful to operations personnel. The proposed contractor for the PLL program should address these issues. For the program to be useful all stakeholders need to buy into its use and structure.

Finding 9. Gaussian puff/plume dispersion modeling techniques embedded in the D2PC computer model used to predict the extent of an agent emission plume are representative of the state of the art as of the late 1970s. Adoption of more modern and accurate emission plume models seems to have been delayed by the failure to integrate better plume models into standard Chemical Stockpile Emergency Preparedness Program emergency response models.

Recommendation 9a. Stockpile sites that still use the D2PC computer model should, at a minimum, upgrade their emergency response models to take advantage of the improved capabilities available in the D2-Puff model. Consideration should be given to testing and possibly optimizing the D2-Puff model at each site by performing tracer release experiments under a variety of meteorological conditions.

Recommendation 9b. The Chemical Stockpile Emergency Preparedness Program should undertake a continuing evaluation of alternative approaches to modeling the release and impact of chemical agents.

Recommendation 9c. Accurate agent plume dispersion modeling capability should be coupled with timely communication of results and appropriate responses to the stockpile site and surrounding communities.

Finding 10a. Communications during and after chemical events have not always occurred as intended between and among the various stakeholders. The lack of a call-forwarding mechanism for getting information directly to people or a hot line dedicated to notification that an event has occurred has contributed to an inadequate communication process during chemical events. The lack of notification and warning between the Deseret Chemical Depot (DCD), Tooele County, and other Utah responsible agencies reflects in part a lack of coordination between components of the two programs (Federal Emergency Management Agency / Chemical Stockpile Emergency Preparedness Program and the Army's Emergency Operations Centers) and in part the DCD's perspective that its emergency management responsibilities "end at the fence." This perspective, if carried to other communities where chemical demilitarization facilities are to be operated, can endanger an effective coordinated emergency response to incidents. The memorandum of understanding recently agreed to by the DCD and Tooele County (Appendix G) for information exchange could serve as a model for every community with a chemical weapons stockpile, to ensure very close oversight of the disposal plant operations.

Finding 10b. The Army's recent and sincere effort to build public trust in its actions has not been sufficiently successful, although the degree of trust or mistrust has not been effectively measured. Of equal or greater importance is public trust in the governmental agencies that monitor the Army's activities. It is essential that these agencies be seen by the public as being fully cognizant of the Army's actions and of being, in effect, in command—a result that will require an extraordinary level of communication between the Army and relevant government oversight agencies and can lead to contradictory advice and requirements that will have to be resolved.

Recommendation 10a. Chemical demilitarization facilities should develop site-specific chemical event reporting procedures and an accompanying training program that tests and improves the implemented procedures and communication system.

Recommendation 10b. The standing incident review board recommended by the committee for each site should include a qualified member of the public who can effectively represent and communicate public interests.

Recommendation 10c. Each chemical demilitarization site should consider the establishment of a reporting and communication memorandum of understanding (MOU), of the sort developed between the Deseret Chemical Depot and Tooele County, which specifies reliable and trusted means of alerting and informing local officials about chemical events. These MOUs should be designed to permit ready evaluation and updating of the terms of the MOU to take full advantage of learning across the array of chemical demilitarization sites.

Recommendation 10d. The Army Emergency Operations Centers and the Chemical Stockpile Emergency Preparedness Program should establish a stronger capability and capacity for the coordination of training, equipment, and plans necessary to respond effectively to an emergency incident, and the commitment to do so in a coordinated and cooperative fashion. Additionally, the Army should continue its program of outreach—including listening to public concerns and responding to them, as well as engaging in more conventional public information efforts—to both the public and the relevant government oversight agencies to enhance general understanding of the chemical demilitarization program.

Finding 11. A major chemical event can result in several months of lost processing time at chemical demilitarization plants. This delays the destruction of the chemical agents, requiring that they remain in the stockpiles where they could create a hazard. When incidents have led to plant shutdown, multiple investigations and responses have contributed to additional delays in restarting operations. All aspects of chemical incident investigations and resumption of operations should be accelerated, consistent with safe operations.

Recommendation 11. All stakeholders and involved regulatory agencies should agree that a single team will investigate chemical events requiring outside review. This investigation team should comprise already-appointed representatives from all stakeholder groups and agencies, including members of the public who can effectively represent and communicate with local officials and the affected public. Incident findings should be documented in a single comprehensive report that incorporates the findings, proposed corrective actions, and concerns of the various oversight agencies.

Finding 12. Safety programs and performance appear to be adequate to ensure that chemical demilitarization operations are being conducted safely. Even so, there is considerable opportunity for improvement. Many of the incidents at Johnston Atoll Chemical Agent Disposal System (JACADS) and Tooele Chemical Disposal Facility (TOCDF) could have been significantly mitigated—if not prevented—had a true "safety culture" been in place and functioning.

Recommendation 12a. Much of the needed improvement in safety at chemical weapons facilities can come from increased attention to factors that contribute to and/or cause chemical events. For example, the Program Manager for Chemical Demilitarization and chemical demilitarization facility managers should ensure that standard operating procedures are in place, up to date, and effective, performing hazard operations analyses on new process steps and design changes even when such changes are viewed as trivial and recognizing that chemical hazards are posed by things other than agent (e.g., waste).

Recommendation 12b. Management at the Tooele Chemical Agent Disposal Facility (TOCDF) and the new third-generation facilities should develop or identify and implement programs that will result in the establishment of a pervasive, functioning safety culture as well as improved safety performance. In doing so, TOCDF and the new chemical demilitarization sites should draw on experience in the chemical industry, obtained through industry associations or other appropriate venues. The Army should revise the award fee criteria to encourage each new chemical demilitarization site operator to demonstrate better safety performance than that at the older sites.

Finding 13. It is probable that conditions will arise in plant operation for which no standard operating procedure has been written. Operators need an in-depth knowledge of their equipment and its limitations to handle these unusual conditions and maintain plant security. New plant start-up can be a difficult learning experience for new operating crews. They need to know how and why procedures are to be performed. It is common practice in other industries to include engineers with "design" knowledge and experience in the start-up crew for new plants.

Recommendation 13. A generous allotment of time should be given to training and retraining chemical demilitarization plant operating personnel to ensure their total familiarity with the system and its engineering limitations. All plant personnel should receive some education on the total plant operation, not just the area of their own special responsibility. The extent of this overall training will be a matter of judgment for plant management, but the training should focus on how an individual's activities affect the integrated plant and its operational risk. Each facility should develop training programs using the newly designed in-plant simulators to present challenges that require knowledge-based thinking. The training programs should include a process for judging the effectiveness of the training. Including "design" experts in the start-up crew for new plants could be helpful in identifying latent failures in process and facility design.

References

SOURCES CITED

Blewett, W.K., D.W. Reeves, V.J. Arca, D.P. Fatkin, and B.D. Cannon. 1996. Expedient Sheltering in Place: An Evaluation for the Chemical Stockpile Emergency Preparedness Program, Chemical Research, Development, and Engineering Center, ERDEC-TR-336. Aberdeen Proving Ground, MD: U.S. Army Armament Munitions Chemical Command.

CDC (Centers for Disease Control and Prevention). 2000. Technical Investigation Report: Release of GB at the Tooele Chemical Agent Disposal Facility (TOCDF) on May 8-9, 2000, June. Atlanta, GA: National Center for Environmental Health.

CSEPP (Chemical Stockpile Emergency Preparedness Program). 2000. Department of the Army, Federal Emergency Management Agency, Chemical Stockpile Emergency Preparedness Program, CESPP Policy Paper Number 18, CSEPP National Benchmarks, August. Washington, DC: Federal Emergency Management Agency.

DoD (Department of Defense). 1980. Methodology for Chemical Hazard Protection, Technical Paper No. 10, Change 3. Alexandria, VA: Department of Defense Explosives Safety Board.

EG&G (Edgerton, Germerhausen and Grier). 2000. Confirmed GB Agent Readings in the Common Stack, Occurrence Report 00-05-08-A1, June 22. Aberdeen Proving Ground, MD: Program Manager for Chemical Demilitarization.

EPA (Environmental Protection Agency). 2001. RCRA Investigation, Johnston Atoll Chemical Agent Disposal System (JACADS), May 9. San Francisco, CA: Environmental Protection Agency Region 9 Office.

FEMA (Federal Emergency Management Agency). 1993. National CSEPP Benchmarks. Washington, DC: Federal Emergency Management Agency.

FEMA. 1996. Planning Guidance for the Chemical Stockpile Emergency Preparedness Program, May 17. Washington, DC: Federal Emergency Management Agency.

FEMA. 1997. Memorandum of Understanding Between the Department of the Army and the Federal Emergency Management Agency: Chemical Stockpile Emergency Preparedness Program (CSEPP), October 8. Washington, DC: Federal Emergency Management Agency.

GAO (General Accounting Office). 2001. Chemical Weapons: FEMA and Army Must Be Proactive in Preparing States for Emergencies, August. Washington, DC: General Accounting Office.

Gertman, D.I., and H.S. Blackman. 1994. Human Reliability and Safety Analysis Data Handbook. New York, NY: Wiley.

Hollnagel, E. 1998. Cognitive Reliability and Error Analysis Method: CREAM. New York, NY: Elsevier.

IEM (Innovative Emergency Management). 2001a. D2-Puff Version 4.0 Reference Manual. Baton Rouge, LA: Innovative Emergency Management, Inc.

IEM. 2001b. D2-Puff Version 4.0 Technical Manual. Baton Rouge, LA: Innovative Emergency Management, Inc.

IOM (Institute of Medicine). 1999. Chemical and Biological Terrorism: Research and Development to Improve Civilian Medical Response. Washington, DC: National Academy Press.

IOM. 2000. To Err Is Human: Building a Safer Health System. Washington, DC: National Academy Press.

Kasperson, R. 1992. The social amplification of risk: Progress in developing an integrated framework. Pp. 153–178 in Social Theories of Risk. S. Krimsky and D. Golding, eds. Westport, CT: Praeger.

Kasperson, R.E., O. Renn, P. Slovic, H.S. Brown, J. Emel, R. Gobel, J.X. Kasperson, and S. Ratnick. 1988. The social amplification of risk: A conceptual framework. Risk Analysis 8(2): 177–187.

NRC (National Research Council). 1994. Review of Monitoring Activities Within the Army Chemical Stockpile Disposal Program. Washington, DC: National Academy Press.

NRC. 1996. Review of Systemization of the Tooele Chemical Agent Disposal Facility. Washington, DC: National Academy Press.

NRC. 1997. Risk Assessment and Management at Deseret Chemical Depot and the Tooele Chemical Agent Disposal Facility. Washington, DC: National Academy Press.

NRC. 1999a. Tooele Chemical Agent Disposal Facility: Update on National Research Council Recommendations. Washington, DC: National Academy Press.

NRC. 1999b. Review and Evaluation of Alternative Technologies for Demilitarization of Assembled Chemical Weapons. Washington, DC: National Academy Press.

NRC. 2000a. Integrated Design of Alternative Technologies for Bulk-Only Chemical Agent Disposal Facilities. Washington, DC: National Academy Press.

NRC. 2000b. Letter Report. A Review of the Army's Public Affairs Efforts in Support of the Chemical Stockpile Disposal Program. Washington, DC: National Academy Press.

NRC. 2000c. Waste Incineration and Public Health. Washington, DC: National Academy Press.

NRC. 2001a. Occupational Health and Workplace Monitoring at Chemical Agent Disposal Facilities. Washington, DC: National Academy Press.

NRC. 2001b. Standing Operating Procedures for Developing Acute Exposure Guideline Levels for Hazardous Chemicals. Washington, DC: National Academy Press.

NRC. 2002. Closure and Johnston Atoll Chemical Agent Disposal System (JACADS). Washington, DC: National Academy Press.

Reason, J. 1997. Managing the Risks of Organizational Accidents. Brookfield, VT: Ashgate.

Scholz, J.T., and F.H. Wei. 1986. Regulatory enforcement in a federalist system. American Political Science Review 80: 1249–1270.

Seinfeld, J.H., and S.N. Pandis. 1998. Atmospheric Chemistry and Physics: From Air Pollution to Climate Change. New York, NY: Wiley.

U.S. Army. 1990. Occupational Health Guidelines for the Evaluation and Control of Occupational Exposure to Nerve Agents GA, GB, GD and VX, Pamphlet 40-8, December. Washington, DC: U.S. Army Medical Services.

U.S. Army. 1991. Occupational Health Guidelines for the Evaluation and Control of Occupational Exposure to Mustard Agents H, HD and HT, Pamphlet 40-173, August. Washington, DC: U.S. Army Medical Services.

U.S. Army. 1993-1995. TOCDF Functional Analysis Workbook. Aberdeen Proving Ground, MD: Program Manager for Chemical Demilitarization.

U.S. Army. 1995. Army Regulation 50-6—Chemical Surety. Washington, DC: Headquarters, Department of the Army.

U.S. Army. 1996a. Tooele Chemical Agent Disposal Facility Quantitative Risk Assessment, SAIC-96/2600. Aberdeen Proving Ground, MD: Program Manager for Chemical Demilitarization.

U.S. Army. 1996b. PMCD Regulation 385-3, Accident and Chemical Event Notification, Investigation, Reporting, and Records, August. Aberdeen Proving Ground, MD: Program Manager for Chemical Demilitarization.

U.S. Army. 1996c. Chemical Agent Disposal Facility Risk Management Program Requirements. Aberdeen Proving Ground, MD: Program Manager for Chemical Demilitarization.

U.S. Army. 1997a. Monitoring Concept Plan. Revision 3, May 1997. Aberdeen Proving Ground, MD: U.S. Army Program Manager for Chemical Demilitarization.

U.S. Army. 1997b. A Guide to Risk Management Policy and Activities, 176-009, January. Aberdeen Proving Ground, MD: Program Manager for Chemical Demilitarization.

U.S. Army. 1999a. Regulation 385-3—Accident and Chemical Event Notification, Investigation, Reporting and Record Keeping, April. Aberdeen Proving Ground, MD: Program Manager for Chemical Demilitarization.

U.S. Army. 1999b. Results of the Independent Verification and Validation Study for the D2-Puff Model, TECOM Project No. 8-CO-480-CSE-001, DPG Document No. WDTC-DR-99-050. Dugway Proving Ground, UT: Metrology and Obscurants Division, West Deseret Test Center.

U.S. Army. 2000a. Annual Status Report on the Disposal of Chemical Weapons and Materiel for Fiscal Year 2000, September 30. Aberdeen Proving Ground, MD: Program Manager for Chemical Demilitarization.

U.S. Army. 2000b. Informal 15-6 Investigation of the Tooele Chemical Agent Disposal Facility (TOCDF) Common Stack Release 8-9 May 2000, June. Aberdeen Proving Ground, MD: Program Manager for Chemical Demilitarization.

U.S. Army. 2000c. Operations Schedule Task Force 2000 Final Report. Aberdeen Proving Ground, MD: Program Manager for Chemical Demilitarization.

U.S. Army. 2001a. Status of Agent Destruction at JACADS and TOCDF, 5 September. Aberdeen Proving Ground, MD: Program Manager for Chemical Demilitarization.

U.S. Army. 2001b. U.S. Army Chemical Demilitarization Program Releases Updated Official Schedule and Cost Estimates, Press Release, October 4. Aberdeen Proving Ground, MD: Program Manager for Chemical Demilitarization.

U.S. Army. 2001c. Information provided to the Committee on the Evaluation of Chemical Events at Army Chemical Agent Disposal Facilities by Andrew P. Roach, Office of the Program Manager for Chemical Demilitarization, May 22.

U.S. Army. 2001d. Information provided to the Committee on the Evaluation of Chemical Events at Army Chemical Agent Disposal Facilities by Gregory St. Pierre, Office of the Program Manager for Chemical Demilitarization, May 30.

U.S. Army. 2001e. Stockpile leak data, provided by M.J. Civis and W.T. Studdert, U.S. Army Soldier and Biological Chemical Command, to the Committee on the Evaluation of Chemical Events at Army Chemical Agent Disposal Facilities, August.

U.S. Army. 2001f. Johnston Atoll Chemical Agent Disposal System: Report of the 3 December 2000 Chemical Agent Reading in the Heated Discharge Conveyor (HDC) Bin, March 30. Aberdeen Proving Ground, MD: Program Manager for Chemical Demilitarization.

U.S. Army. 2001g. Information provided to the Committee on the Evaluation of Chemical Events at Army Chemical Agent Disposal Facilities during a fact-finding trip, Tooele Chemical Agent Disposal Facility, Tooele, UT, July 26.

U.S. Army. 2002a. Information provided to the Committee on the Evaluation of Chemical Events at Army Chemical Agent Disposal Facilities by Conrad Whyne, Deputy Project Manager for Chemical Stockpile Disposal, September 12.

U.S. Army. 2002b. Information provided to the Committee on the Evaluation of Chemical Events at Army Chemical Agent Disposal Facilities by the Program Manager for Chemical Demilitarization, March 5.

USCNS (United States Commission on National Security). 2001. Roadmap for National Security: Imperative for Change, January 31. Available online at <http://www.homelandsecurity.org/sugg_reading/Phase_III_Report.pdf> [July 11, 2002].

USNRC (U.S. Nuclear Regulatory Commission). 2000. Technical Basis and Implementation Guidelines for a Technique for Human Event Analysis (ATHEANA), NUREG-1624, REV. 1. Rockville, MD: U.S. Nuclear Regulatory Commission.

Utah DEQ (Department of Environmental Quality). 1996. Tooele Chemical Agent Disposal Facility Screening Risk Assessment, EPA I.D. No. UT5210090002. Salt Lake City, UT: Utah DEQ.

Utah DEQ. 2000a. Investigation Report on the Agent Release from the Common Incinerator Stack on May 8 and 9, 2000 at the Tooele Chemical Agent Demilitarization Facility, June 16. Salt Lake City, UT: Utah DEQ Division of Solid and Hazardous Waste.

Utah DEQ. 2000b. Memorandum of Understanding Between Deseret Chemical Depot and Tooele County for Information Exchange, September 12; updated November 7, 2001. Salt Lake City, UT: Utah DEQ Division of Solid and Hazardous Waste.

Weick, K.E., and K.M. Sutcliffe. 2001. Managing the Unexpected: Assuring High Performance in an Age of Complexity. San Francisco, CA: Jossey-Bass.

Wood, B.D. 1992. Modeling federal implementation as a system: The clean air case. American Journal of Political Science 36: 40–67.

LAWS GOVERNING CHEMICAL DEMILITARIZATION

Treaty

Convention on the Prohibition of the Development, Production, Stockpiling and Use of Chemical Weapons and on Their Destruction (P.L. 105-277)

Statutes

Chemical Safety Information, Site Security and Fuels Regulatory Relief Act (P.L. 106-40)
Clean Air Act (CAA; 42 U.S.C. §7401 et seq.)
Clean Water Act (CWA; 33 U.S.C. §1251 et seq.)
Emergency Planning and Community Right to Know Act (EPCRTKA; 42 U.S.C. §11001 et seq.)
Occupational Safety and Health Act (OSHA; 29 U.S.C. 1920.120 et seq.)
Resource Conservation and Recovery Act (RCRA; 42 U.S.C. §6901 et seq.)
Toxic Substance Control Act (TSCA; 15 U.S.C. §2601 et seq.)

APPENDIXES

Appendix A

Specific Design Features of the Toole Chemical Agent Disposal Facility Baseline Incineration System

PROCESS DESCRIPTION

The Tooele Chemical Agent Disposal Facility (TOCDF) consists of five interconnected process systems:

1. The unloading and unpack system for receiving munitions from the Deseret Chemical Depot.
2. The demilitarization processing systems for handling rockets, containers, mines, and projectiles separately.
3. The furnace and incinerator systems, which include a deactivation furnace system, a metal parts furnace, two liquid incinerators, and a dunnage incinerator.
4. Various safety systems, including explosive containment, ventilation and filtering, fire protection, agent monitoring, and door monitoring.
5. Various support systems, including electric, fuel gas, instrumentation, compressed air, hydraulics, cooling, and the very important pollution abatement systems.

These systems are linked, monitored, and controlled through an advanced process management system operated from a central control room.

For practical purposes, the TOCDF is a scaled up and updated version of the Johnston Atoll Chemical Agent Disposal System (JACADS), which has been operating for nine years. Although JACADS was the first chemical agent disposal facility, its design was based on pre-existing commercial incinerators, as well as years of development and testing of special munitions-handling machinery. Very little new technology was incorporated into the TOCDF. The layout of the TOCDF is shown in Figure A-1.

Unloading and Unpack System

Munitions are brought by truck in sealed containers from the storage area in Deseret Chemical Depot into the container-handling building along dedicated and highly secure roads. The containers are lifted to the second floor of the building into the unpack area where they are opened, and the munitions are conveyed into the munition demilitarization building. No human contact with the munitions occurs after the munitions leave the unpack area.

Demilitarization Processing Systems

The purpose of demilitarization processing is to separate the components of munitions into separate streams that can be handled safely in the downstream furnace and incinerator systems. Each type of munition is unique and must be processed separately. Rockets, for example, contain agent, propellant, and burster energetics, which must be separated for processing. The rocket-handling system feeds rockets into an explosion-containment room through a rotating vestibule. In the explosion-containment room, the agent cavity is punched open, and the agent is drained into a separate holding tank. Eventually, the agent is fed into a liquid incinerator (LIC) and burned. The drained rocket proceeds to a shearing device where the fuse is sheared off, the burster is sheared off, and finally the propellant-containing motor is sheared off. The fuse, burster, and motor fall into a hopper that discharges them into the deactivation furnace system (DFS). The rocket-handling system is shown in Figure A-2.

Bulk munitions contain agent but no energetics. Therefore, they bypass the explosion-containment room and are conveyed into the upper munitions corridor of the munitions processing building to a bulk drain station. Bulk containers are hydraulically punched so that agent can be drained into a holding tank prior to incineration in a LIC. The drained container and the tray it was on are conveyed to the metal parts furnace (MPF) for cleanup. The bulk handling system is shown in Figure A-3.

NOTE: This appendix is reprinted from NRC (1999a), pp. 61-70.

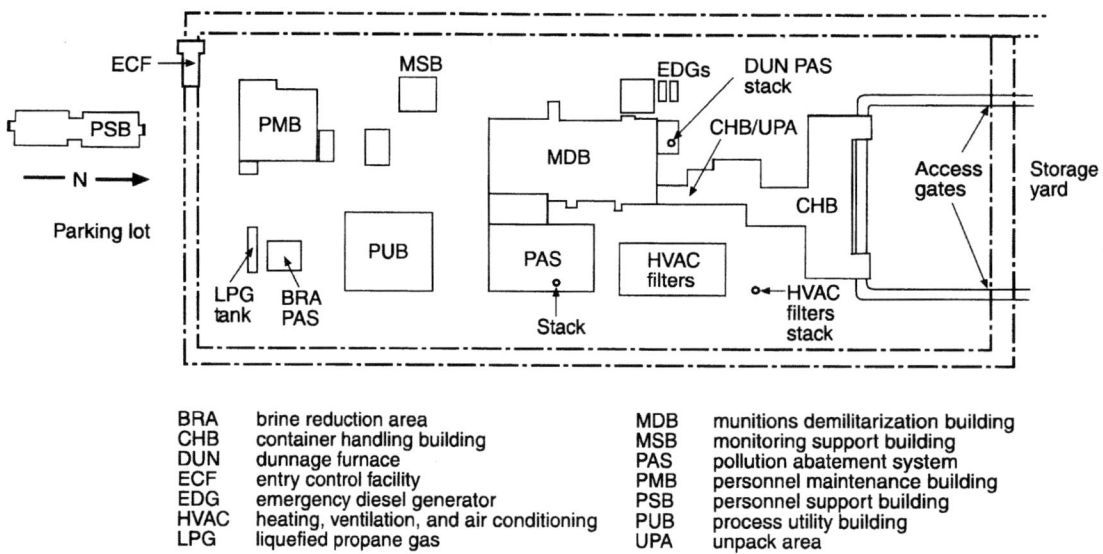

FIGURE A-1 Layout of the TOCDF. Source: Adapted from U.S. Army, 1996.

Projectiles are processed in a system similar to the rocket-handling system. Projectiles, either artillery shells or mortar shells, contain both agent and energetics. Projectiles enter the explosion-containment room by conveyer and are fed mechanically onto a projectile/mortar disassembly table. The table rotates so that nose closures (fuses or lifting lugs) can be mechanically removed. At another stop, burster material is removed. The shells are then placed in an egg-crate metal tray and conveyed into the munitions processing bay located in the upper munitions corridor. A robot unloads the shells onto another rotating table called the multipurpose demilitarization machine, where they are milled to cut through burster tube welds, if necessary. Then the burster tubes are removed, and the agent is drained. Finally, the burster tube is crimped and reinserted, and the projectile is sent through the MPF. The projectile-handling system is shown in Figure A-4.

The mine-handling system is the last demilitarization processing system. Operators unpack mines from their drum containers in the unpack area. Each mine is then cycled through a glove box onto a conveyer in the explosion-containment vestibule. This conveyer takes them to a workstation where the arming plugs, fuses, and activators are removed and placed in a fuse box. The fuse box and the mine are then transported to the explosion-containment room, where a mine machine punches the mine and drains the agent. A burster punch machine removes the burster from the mine. The remnants of the mine and the fuse box are then sent to the DFS. Figure A-5 depicts the mine-handling system.

Furnaces and Incinerators

The DFS is used to destroy explosives and propellants from rockets, projectiles, and mines. Basically, the DFS is a gas-fired rotary kiln (Figure A-6). Munitions pieces are fed down a chute from the explosion-containment room into the DFS. The chute has two blast gates that open sequentially. As the kiln rotates, the pieces are moved through the kiln by a spiral baffle that pushes them along. For rocket campaigns, the kiln runs at 1,100°F. For other campaigns, it runs at 1,500°F. The pieces burn rapidly rather than detonating. As added protection against detonation, the charge end of the kiln is constructed of two-inch thick steel. The burned munitions exit onto a discharge conveyer that carries them under two electric heater banks that keep the scrap at 1,000°F for 15 minutes. This ensures that the scrap is 5X clean, (i.e., 99.99999 percent free of agent). DFS exhaust gases go through a blast-attenuation duct, a cyclone separator (to remove ash), and an afterburner before entering the pollution abatement system (PAS).

The function of the MPF is to decontaminate munitions bodies after removal of agent and explosives. The

APPENDIX A

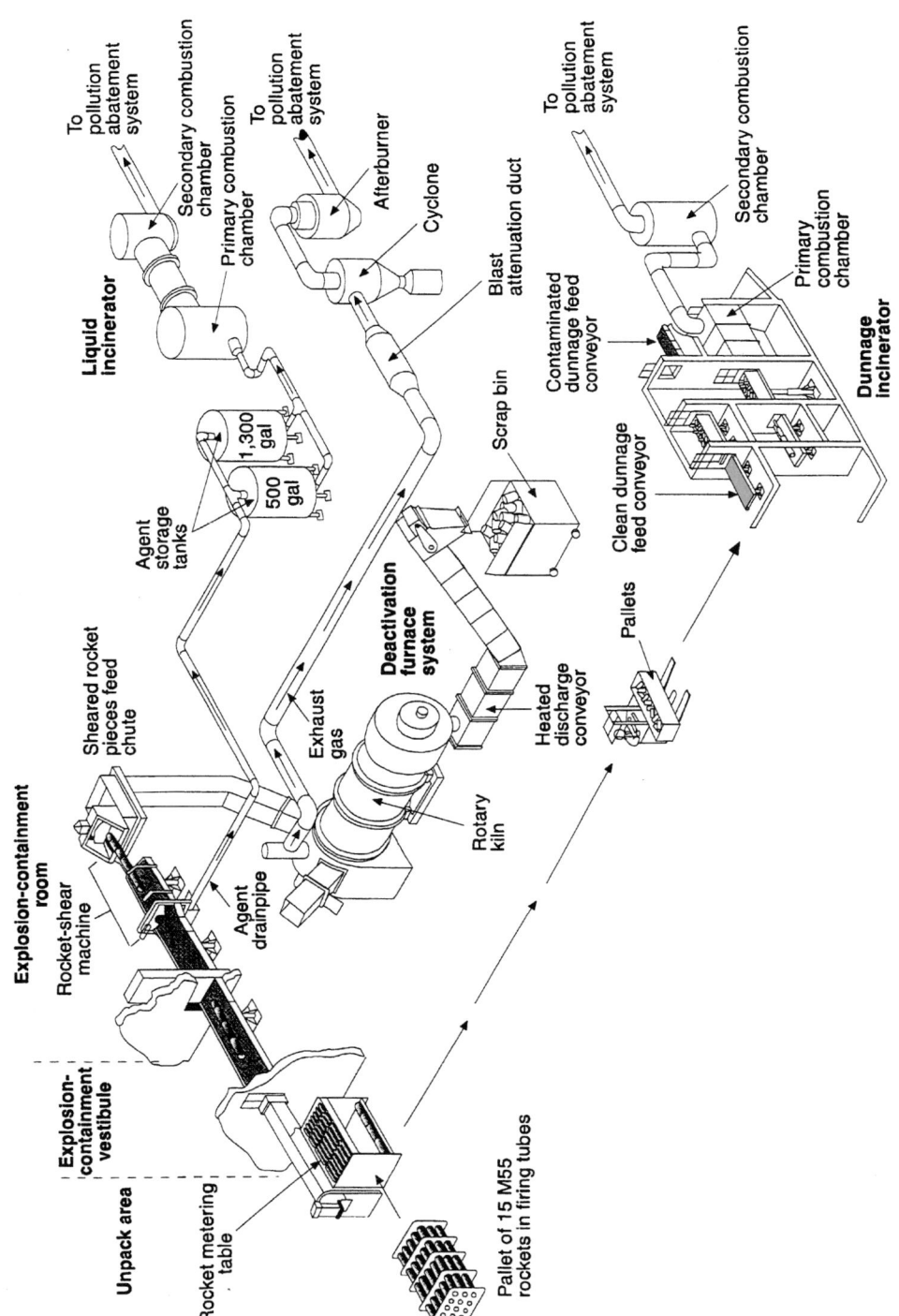

FIGURE A-2 Rocket-handling system. Source: Adapted from U.S. Army, 1996.

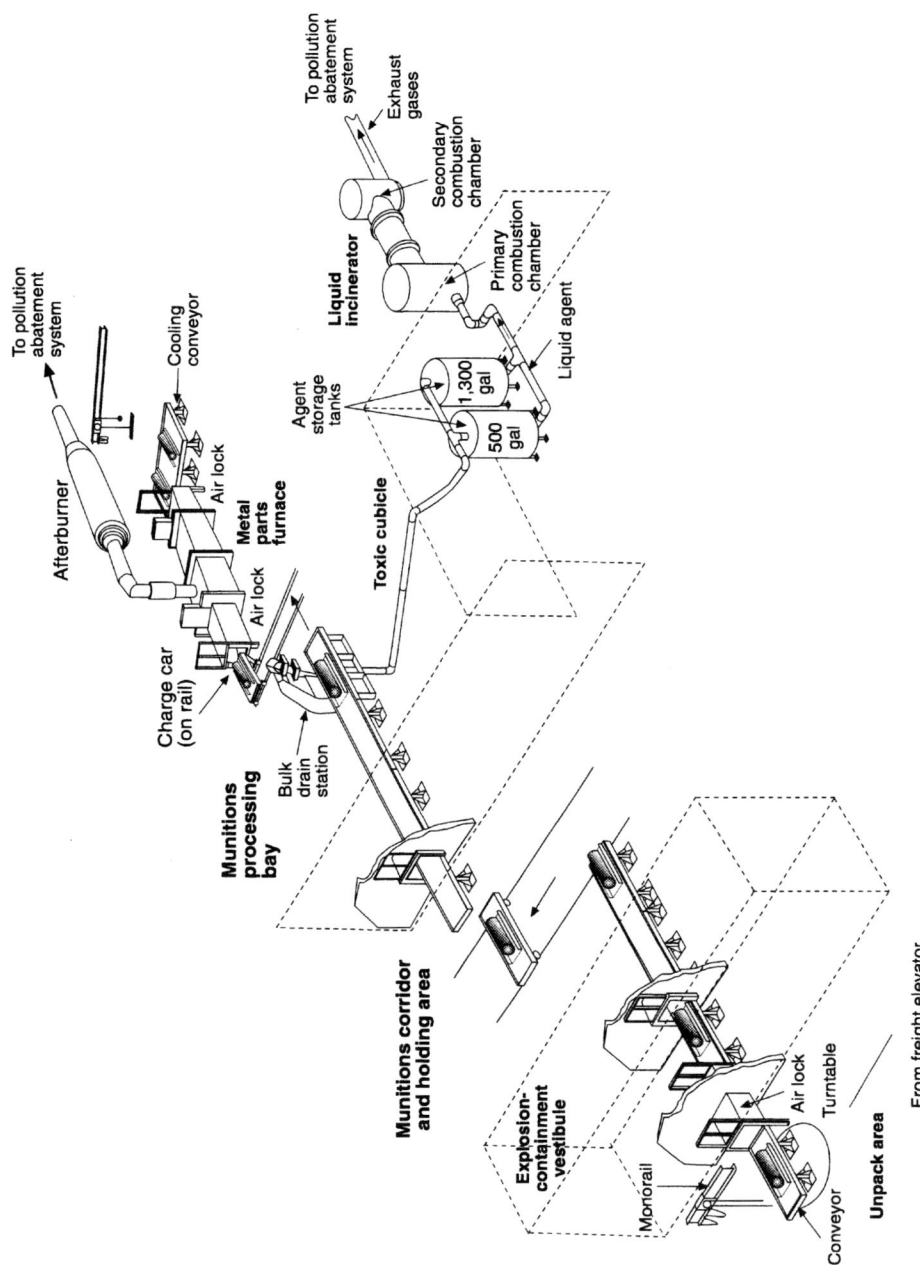

FIGURE A-3 Bulk handling system. Source: Adapted from U.S. Army, 1996.

APPENDIX A

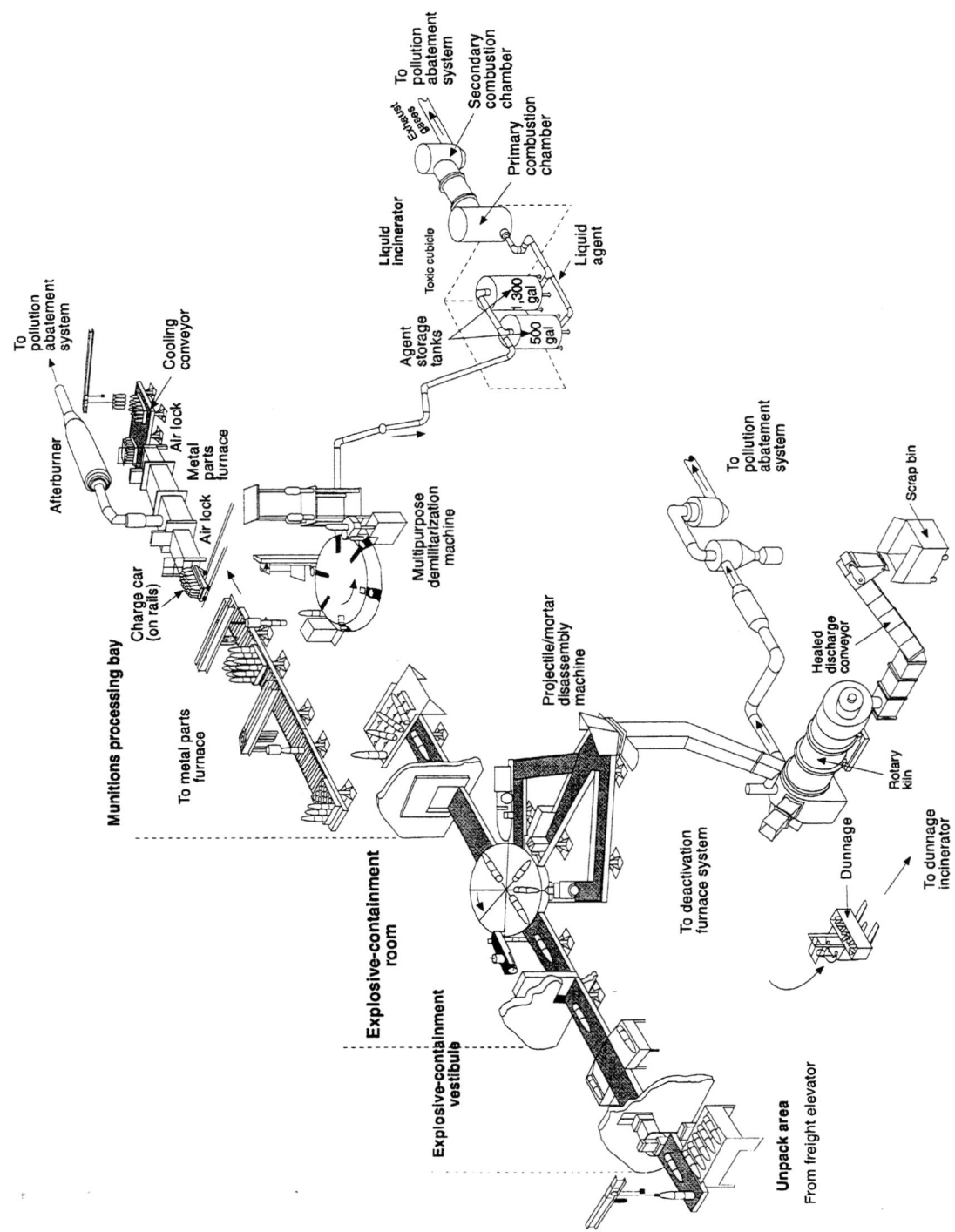

FIGURE A-4 Projectile-handling system. Source: Adapted from U.S. Army, 1996.

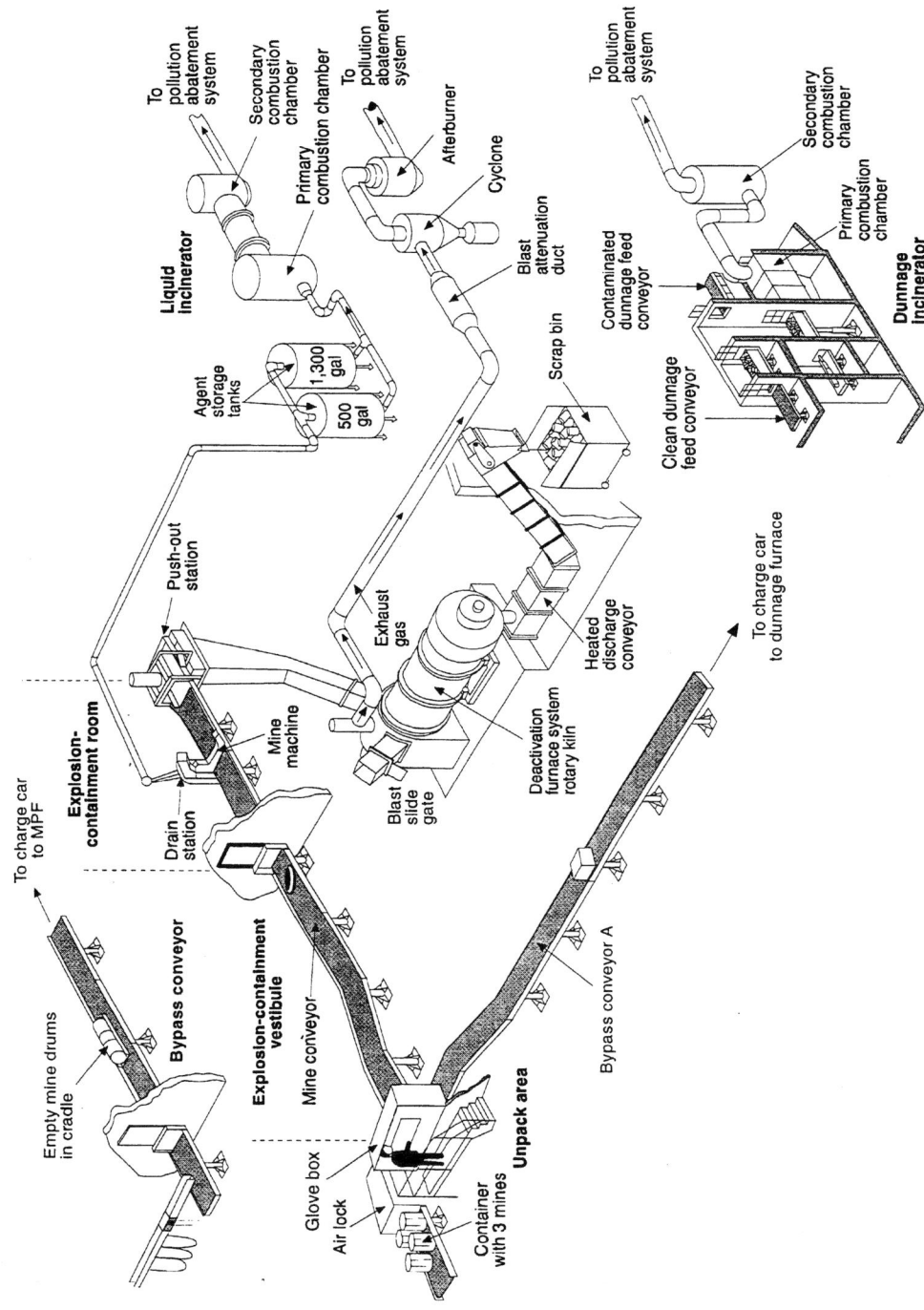

FIGURE A-5 Mine-handling system. Source: Adapted from U.S. Army, 1996.

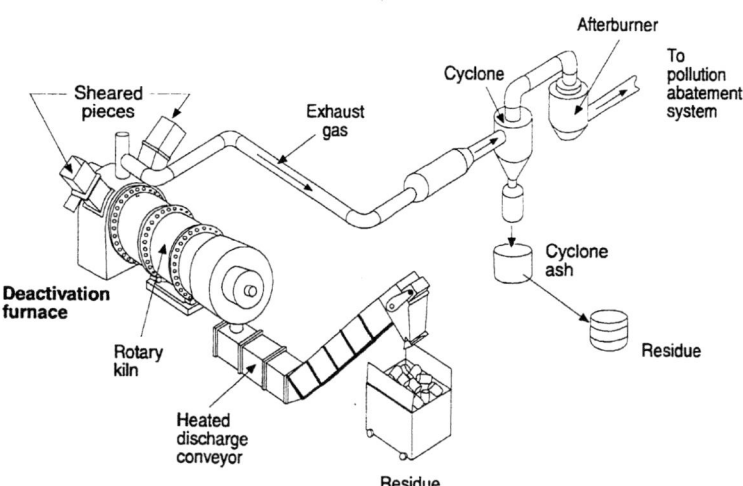

FIGURE A-6 Deactivation furnace system. Source: Adapted from U.S. Army, 1996.

MPF is diagrammed in Figure A-7. For ton containers, the MPF peaks at 1,450°F. For spray tanks, it operates at 1,525°F. For smaller items, it operates at 1,600°F. Contaminated items are conveyed semicontinuously through a charge air lock into the first of three heating zones, each of which has an air-lock door. Pieces are held in the discharge air lock until they cool enough so that agent levels can be monitored. Pieces that are 5X clean are cooled and containerized for disposal. The exhaust gas from the MPF goes through an afterburner and then to the PAS.

Two LICs destroy liquid agent. Figure A-8 shows the LIC configuration. The primary chamber, a vertical refractory-lined cylinder with a natural gas burner, operates at 2,700°F. Agent is atomized as it is injected into the air stream going into the burner. As the agent burns,

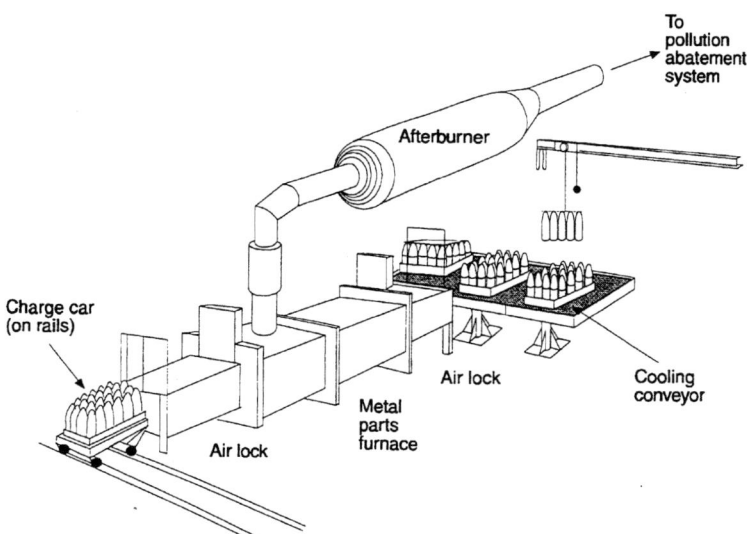

FIGURE A-7 Metal parts furnace. Source: Adapted from U.S. Army, 1996.

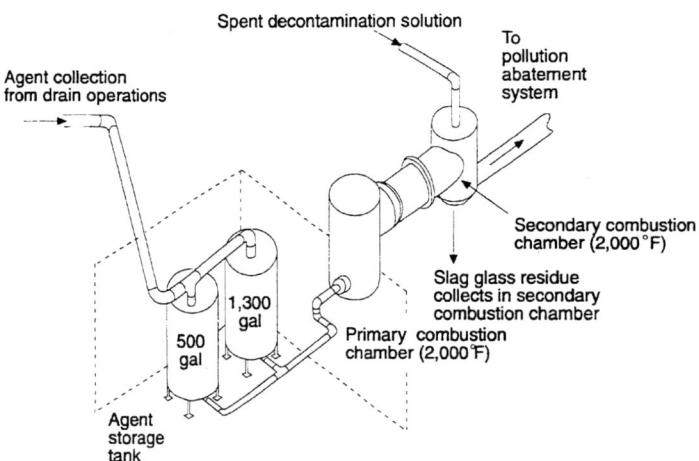

FIGURE A-8 Liquid incinerator. Source: Adapted from U.S. Army, 1996.

the natural gas supply is cut back to maintain the temperature at the desired level. The exhaust from the primary chamber goes into a similar, refractory-lined secondary chamber, in which the temperature is maintained at 2,050°F by burning natural gas. Spent decontamination solution is atomized and injected into the second combustion chamber. All of this forms a molten slag, which is drawn off through a bottom tap into barrels, where it solidifies. Once cool, these barrels are covered and stored prior to disposal.

A dunnage incinerator (DUN) is designed to destroy the plastic, wood, or paper packing cases, pallets, and other objects that may be contaminated by agent. In practice at the TOCDF, the DUN has not operated routinely because the listed materials could be safely disposed of in other ways. The DUN is designed to burn natural gas and dunnage combustibles at a temperature of 1,400°F. The configuration of the DUN is shown in Figure A-9. The primary combustion chamber is refractory-lined and has four side burners. Air is supplied both through

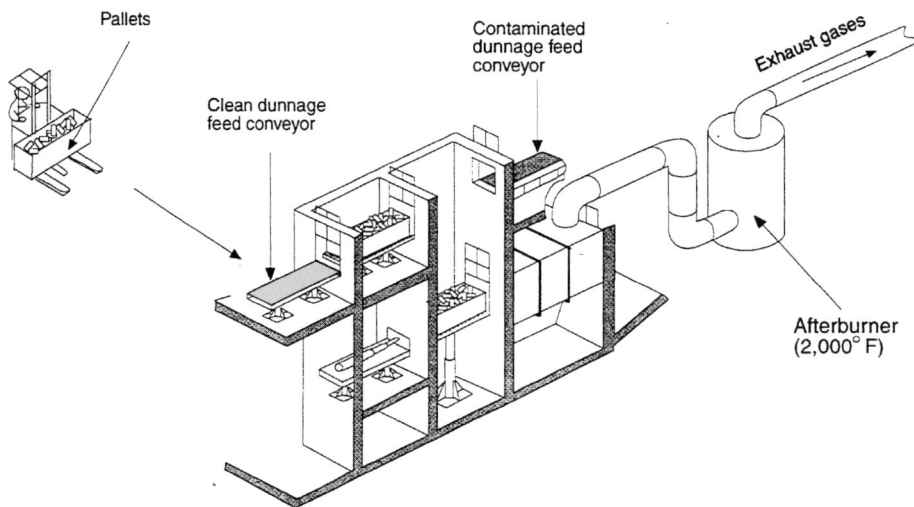

FIGURE A-9 Dunnage furnace. Source: Adapted from U.S. Army, 1996.

APPENDIX A

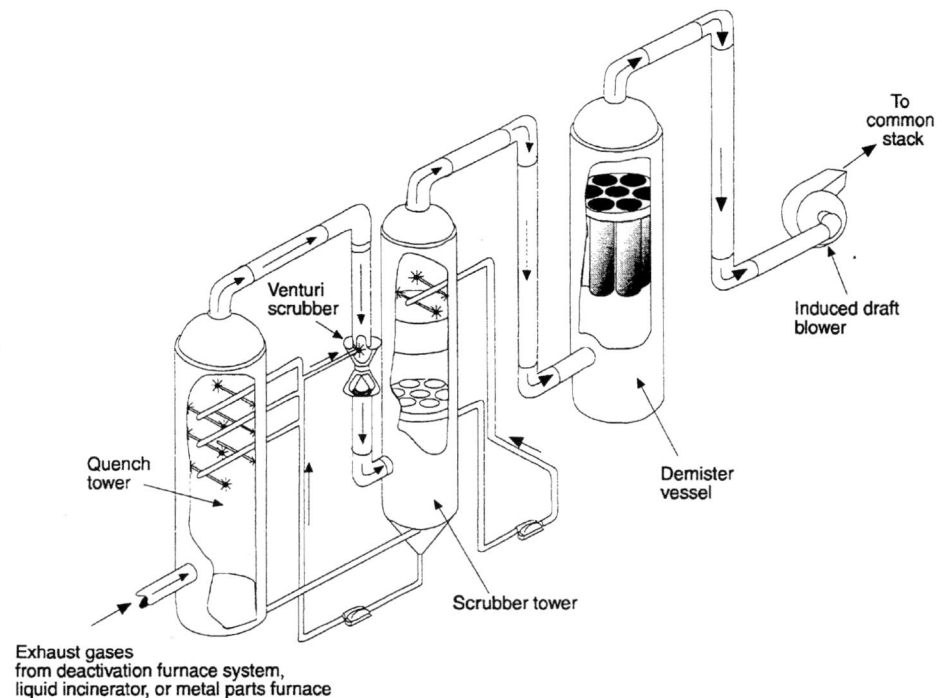

FIGURE A-10 Pollution abatement system. Source: Adapted from U.S. Army, 1996.

the burners and through side wall ports. Ashes are removed from the furnace periodically. Exhaust gases go to the afterburner, which operates at 2,000°F. Secondary exhaust passes into the PAS.

Safety Systems

Ensuring process safety is the prime concern of the design and operation of the TOCDF. Explosion-containment requirements were mentioned in several of the preceding sections. The overall design for explosion-containment rooms requires containment of a blast from 15 pounds of TNT. The DFS room is designed to contain a blast from 28.2 pounds of TNT. Interlocked blast gates and blast doors are used to ensure containment.

Agent dispersion in the air stream is another major safety concern. Avoiding contamination is accomplished by pressure cascading the air flow throughout the plant from areas with low contamination probability through areas with increasing contamination probability. The air from the most susceptible areas to agent contamination (the furnace rooms and the munition demilitarization building) is filtered through a series of high efficiency particulate air filters and carbon adsorption beds before being exhausted to a stack. In situ monitoring for agent occurs at many points within and around the perimeter of the plant. In addition, ambient air is continuously pumped through contaminant concentration tubes that are periodically collected and analyzed for agent by gas chromatography. There is also a system for monitoring and controlling doors so that the ventilation flowpaths are not upset even when personnel enter or leave the munition demilitarization building areas.

Fire protection is another critical safety concern. Automatic fire detectors are located throughout the plant. Sprinkler systems supplied from a large storage tank come on automatically in the event of a fire in the unloading and unpack areas. In other areas, dry chemical systems are deployed. Halon systems protect the control room and power supply room.

Support Systems

The electric, instrumentation, compressed air, hydraulics, fuel gas, and cooling systems are fairly standard industrial systems, but they are often paralleled to ensure reliability. Each furnace system has a downstream PAS to neutralize and remove the acidic components (hydrochloric, hydrofluoric, sulfuric acids, etc.) formed during the combustion of the agent so the exit gas can be safely released to the atmosphere. Figure A-10 illustrates a typical PAS configuration. The furnace outlet gases enter a quench tower in which a caustic solution is sprayed. The cooled gases exit into a venturi scrubber where they are again in contact with caustic brine. Finally, they go through a scrubber tower where they are in contact with additional brine, through an induced draft fan, and then to a common stack. The PAS for the DUN is simple. It has only a quench tower because the exit gases are far less acidic than those from the other furnaces.

The brine reduction area (BRA) process involves evaporating brine with steam generated on site, then drying it to salt with less than 10 percent water content. The gas from the evaporator is superheated and passed through a bag filter system before being exhausted to the atmosphere. Currently, brine from the PAS is collected, stored temporarily, and then disposed of off site as a hazardous waste. This brine disposal strategy is currently a cheaper alternative than operating the BRA.

Operations Control Room

The central control room provides surveillance and direction for all phases of TOCDF activities. It is kept at a higher positive pressure to prevent the possibility of any agent entering it, and the air intake is doubly filtered. Several consoles line the room, each with two advisor screen monitors, two closed-circuit TV monitors, and a keyboard through which commands are entered to control plant operations. Redundant computers, software, and plant instrumentation ensure that continuous real-time control is maintained.

Reference

U.S. Army. 1996. Tooele Chemical Agent Disposal Facility Quantitative Risk Assessment. SAIC-96/2600. Aberdeen Proving Ground, Md.: U.S. Army Program Manager for Chemical Demilitarization.

Appendix B

Chronicle of Chemical Events and Other Occurrences at TOCDF and JACADS As Identified by PMCD

TOCDF

Date	Event Control Number	Building	Comments
24-Aug-96	TOCDF 96-01 (Chemical Event) **[PMCD]**	Carbon Filter Vestibule	GB agent migration from MDB filter into vestibule. Probable cause of the agent readings in the containment vestibules was the method used for placing the filter units off-line when not in service. The method used closed both the inlet and outlet dampers, which allowed the pressure in the filter plenum to equilibrate with the filter containment vestibules. No indication of a leak.
09-Oct-96	TOCDF 96 – (Chemical Event)	Chemical Analysis Lab - Rm. 113	Operators used the room-breathing zone ACAMS to monitor for agent vapors inside the transportation over pack.
14-Oct-96	TOCDF 97-01 (Chemical Event)	Explosive Containment Room B	Rocket remained at the punch and drain station with the sequencer not reset due to PLS 3-sensor failure. Rocket hung up on conveyor. Rocket freed and processed.
26-Jan-97	TOCDF 97-02 (Chemical Event)	Observation Corridor - Buffer Storage Area	Operators had shut down the LIC # 2 and adjusted the atomizing air pressure to stabilize flame strength. Ventilation upset allowed vapor to migrate to the Observation Corridor. Agent maintained under engineering controls.
08-Apr-97	TOCDF 97-03 (Chemical Event)	Chemical Assessment Lab	Waste was being transferred from satellite hood # 5 to the 90-day area hood # 6. Following the transfer, ACAMS went into alarm.

NOTE: This appendix is reprinted from U.S. Army (2001c). Boldfaced bracketed notations in the column headed "Event Control Number" indicate that external investigations were conducted. Legend: CDC, Centers for Disease Control and Prevention; DAIG, Department of the Army Inspector General; DA Safety, Department of the Army Office of Safety; DOD, Department of Defense; PMCD, Program Manager for Chemical Demilitarization; UTAH DEQ, Department of Environmental Quality, State of Utah.

Date	Report	Location	Description
30-May-97	TOCDF / DCD 97-22 (Chemical Event) **[DAIG & DOD]**	CHB	Three (3) MC-1 bomb casings were delivered to TOCDF from DCD. The bomb casings were certified 3X. A low-level vapor reading of GB was detected inside an empty MC-1 bomb casing. ACAMS located within 12 feet of the bomb casings did not detect agent. DAIG conducted investigation of allegations of civilian exposure to chemical agent during a tour of TOCDF. Conclusion of investigation, Army policy does not adequately address protection of civilians in a chemical hazard area. Regulatory guidance and procedures changed.
06-Aug-97	TOCDF 97-04 (Chemical Event)	MBD - Observation Corridors	During scheduled power outage test, back-up power did not come on line in the prescribed time. Agent vapor migration to Observation Corridors due to HVAC failure during global loss of power.
30-Mar-98	TOCDF 98-01 (Chemical Event)	DEMIL FACILITY	Incomplete drain of an MC-1 bomb allowed excess of agent to enter metal parts furnace. The resulting temperature variations caused the system to shut down as designed. No agent was release to the environment as a result of the event.
13-Dec-98	TOCDF 99-01 (Chemical Event)	Toxic Cubicle	The strainer system on ACS-tank-101 was incorrectly re-assembled following solids removal from the strainer basket. This allowed approximately 142 gals of liquid agent to spill into the containment sump of the Toxic Cubicle. The TOX containment sump can hold 512 gals of liquid. The agent was collected into the sump in the toxic cubicle and subsequently pumped to the agent spill tank. The TOX is designed for this contingency. The cause of the leak was an improperly sealed lid following a change-out of the agent strainer.
04-Apr-99	Weekly Report (GM 0674-99) (Chemical Event)	DFS Cyclone Bin	DAAMS tube detected agent for the DFS cyclone bin enclosure at 16.36 TWA. A second DAAMS analysis resulted in 7.99 TWA. A portable ACAMS was then used to monitor the cyclone bin and enclosure. ACAMS readings were 0.0 TWA. Real-time Analysis Platform (RTAP) support was requested to monitor the area around the cyclone enclosure. The results of the RTAP monitoring were 0.00 TWA.
02-May-99	Weekly Report (GM 0756-99)	Unpack Area (Chemical Event)	Unpack Area (UPA) ACAMS - 203 alarmed at maximum reading of 508 TWA and the local ACAMS displayed 2080 TWA. UPA workers were evacuated and monitored out using ACAMS. Blood draws were performed on all potentially exposed workers. Blood draws on workers indicated no depression of CHE. Once the ACAMS level dropped below the level of quantification, Monitoring Personnel went into the UPA to challenge ACAMS-203 and pull DAAMS tubes. ACAMS-203 alarmed at 0.29 TWA while personnel were in level C protective equipment. DAAMS tubes confirmed for GB Agent at 0.54 TWA. A safety investigation conducted.
21-May-99	TOCDF 99-02 (Chemical Event) **[PMCD]**	Unpack Area	Agent migrates from an Explosive Containment Room to Unpack Area (UPA). UPA ACAMS alarmed and seven workers donned their masks. Blood draws on workers indicated no depression of CHE. Probable source of agent vapors was migration from the ECR through the EVC to the UPA through the transfer airlocks, during the transfer of 2 ton containers.

APPENDIX B

Date	Event	Location	Description
24-May-99	TOCDF 99-03 (Chemical Event) **[PMCD]**	Unpack Area	After removal of nose closures on 105mm projectiles, one had liquid agent in the burster well. Agent was not suspected to be present in the burster well.
26-May-99	TOCDF 99-04 (Chemical Event) **[PMCD]**	Toxic Maintenance Area	Workers in the Toxic Maintenance Area (TMA) were removing bags of waste from trays for segregation and further processing. ACAMS alarmed and workers exited the TMA. Blood draws on workers indicated no depression of CHE.
04-Jun-99	TOCDF 99-05 (Chemical Event) **[PMCD]**	MDB	TOCDF experienced a global loss of power due to a loss of a substation in Tooele, Utah. Emergency generator # 101 failed to take the electrical load initially and generator # 102 had frequency and voltage fluctuations. These fluctuations caused breakers to trip open. MDB ventilation system began to degrade allowing agent vapor migration. A technical engineering analysis of the emergency electrical system was conducted.
20-Feb-00	TOCDF 00-01 (Chemical Event)	DFS Room	Two workers were repairing the Heated Discharge Conveyor (HDC) when an ACAMS alarmed. Workers exited immediately and sent to TOCDF Medical Clinic for evaluation. Blood draws on workers indicated no depression of CHE.
23-Feb-00	TOCDF 00-02 (non-surety)	Liquid Incinerator (LIC # 2)	30-gal barrel of slag tipped over by mechanical lift and ignited a small fire, which damaged a slag conveyor motor and electrical controls. Cause was failure of a sensor which stops conveyor when a barrel is present. Repairs to Slag conveyor motor and electrical controls.
27-Feb-00	TOCDF 00-03 (non-surety)	CHB	Agent detected inside an On-Site Container (ONC). The ONC is used to safety transport munitions from the storage area to demil facility. ONC moved to the UPA and processed. Leak contained in engineering controls.
19-Mar-00	TOCDF 00-04 (non-surety)	TOCDF	High winds interrupted commercial power. As a precaution TOCDF suspended munition processing. Demil process resumed after commercial power restored.
20-Apr-00	TOCDF 00-05 (non-surety)	DFS Air lock	Commercial power disrupted and TOCDF changed to emergency backup power. ACAMS alarmed in the DFS due to the interruption of the airflow. As a precaution TOCDF suspended munition processing and RTAPS deployed with negative readings. Demil process resumed after commercial power restored.
08-May-00	TOCDF 00-06 (Chemical Event) **[DA SAFETY, CDC & UTAH DEQ]**	TOCDF	During processing of GB rockets the Deactivation Furnace System (DFS) interlock shut off all burners due to pollution abatement system (PAS) air flow meter failure. ACAMS alarmed in the furnace stack during re-light of the furnace. No agent or munitions were being processed at time of the alarms. The perimeter monitor readings were all negative for agent. Investigation teams from CDC, Department of Army Safety and Utah (DSHW) conducted the investigation of stack release. Technical investigation completed with recommended procedural and design changes.

Date	Event	Location	Description
26-Jun-00	TOCDF 00-07 (non-surety)	TOCDF	Non-confirmed reading - False positive readings at common stack caused by contaminated DAAMS tubes. Contamination caused by the calibration standards. Event closed.
14-Aug-00	TOCDF 00-08 (non-surety)	TOCDF (Non-surety)	Office employee suffering from acute allergies and Asthma had an adverse reaction to the cutting of weeds. The employee was transported to the hospital for treatment. Hospital doctors recommended she remain for observation and released the next day.
13-Oct-00	TOCDF 01-01 (Non-surety)	DCD BLDG. 4544	Truck driver fell from his truck and bruised his shoulder. The driver was taken to the hospital for treatment and released. No agent or munitions involved.
15-Oct-00	TOCDF 01-02 (Non-surety)	TOCDF	Employee had cardiac symptoms and was transported to hospital. Doctors concluded the man suffered from digitalis (excess salt) due to poor diet. Non-work related and no agent or munitions involved.
26-Oct-00	TOCDF 01-03 (Non-surety)	TOCDF	Employee suffered from shortness of breath due to a respiratory infection. Non-work related and no agent or munitions involved. Sent to hospital.
03-Nov-00	TOCDF 01-04 (Non-surety)	TOCDF	A propane leak at TOCDF caused a temporary shut down of operations. DCD Fire Department was summoned. The fire department set up gas detectors and determined that there was no immediate danger. No agent or munitions involved.
17-Nov-00	TOCDF 01-05 (Non-surety)	TOCDF	Operational shutdown - natural gas supplier Questar, informed DCD due to unseasonably cold weather gas supply was low. 20 Nov natural gas supply restored and processing resumed.
25-Nov-01	TOCDF 01-06 (Non-surety)	Personnel Maintenance Building,	A bag of TAP gear was being monitored when ACAMS alarmed. Bag sent to TMA to be processed.
26-Nov-01	TOCDF 01-07 (Non-Surety)	Personnel Maintenance Building,	A bag of TAP gear was being monitored when ACAMS alarmed. Bag sent to TMA to be processed.
05-Dec-00	TOCDF 01-08 (Non-Surety)	Liquid Incinerator (LIC # 1)	Slag barrel spilled causing a small fire. Repairs to Slag conveyor motor and electrical controls.
18-Dec-01	TOCDF 01-09 (Non-Surety)	Filter Bank # 9	During filter change operation the plastic bag ripped. No agent detected.
28-Dec-00	TOCDF 01-10 (Non-Surety)	TOCDF	Employee had symptoms of an appendicitis or gall bladder attack. He was transported hospital. Non-work related and no agent or munitions involved.
01-Jan-01	TOCDF 01-11 (Non-Surety)	ACAMS 107B	TOCDF experienced a small electrical fire (heat trace) with ACAMS sample line. Fire extinguished with dry-chemical fire extinguisher. No agent or munitions involved.
02-Feb-01	TOCDF 01-12 (Non-Surety) **[PMCD]**	Chemical Assessment Lab (CAL) - 90 Day hood (Non-Surety)	Lab waste was not sufficiently deconned prior to movement to the 90-day hood. Lab area under engineering controls no agent release. Local investigation being conducted.

APPENDIX B

Date	Event Control Number	Building	Comments
16-Feb-01	TOCDF 01-13 (Non-Surety) [PMCD]	CAL - Rm. 114	Two 50-ml plastic centrifuge tubes containing 10-ml solution were transported between rooms without secondary containment. No Lab personnel were in room at time of the ACAMS alarm.
21-Feb-01	TOCDF 01-14 (Non-Surety) [PMCD]	CAL - Rm. 115	Waste vial was not filled with bleach. Investigation into Lab practices initiated covering recent Non-surety events with the Lab.
27-Feb-01	TOCDF 01-15 (Non-surety)	Liquid Incinerator-Secondary Room	Inadvertently shut down of ventilation system. All agent vapor maintained under engineering controls. LIC secondary room was maintained under engineering controls. No agent release.
13-Mar-01	TOCDF 01-16 (Non-surety)	Liquid Incinerator (LIC # 2)	Slag barrel overfilled during slag tap causing a small fire. Damaged electrical wiring and conveyor system in the slag pit.
02-Apr-01	TOCDF 01-17 (Non-surety)	Explosive Containment Room B	Bottom gated opened and failed to cycle completely - Gate remained in the open position. Sheared rocket pieces overheated and flashed. Deluge system activated extinguishing the fire.
19-Apr-01	TOCDF 01-18 (Non-surety)	Personnel Maintenance Building	Monitoring of TAP Gear in the cotton goods room ACAMS alarmed. Bags moved to the Toxic Maintenance Area for further processing.
05-May-01	TOCDF 01-19 (Non-surety)	Liquid Incinerator (LIC # 2)	Slag fire of less than 1 pound burned out a sensor. Not reported as event because of size.

JACADS

Date	Event Control Number	Building	Comments
30-Jun-90	JACADS – (Unusual Occurrence) [PMCD]	Munitions Demilitarization Building (MDB)	After processing the first 15 rockets, the DAAMS tubes were collected and analyzed. Several DAAMS tubes were confirmed positive, but there were no ACAMS alarms. Determined to be the analysis of a DAAMS tube from the ECR which was saturated with GB.
08-Dec-90	JACADS – (Unusual Occurrence)	Common Stack	Agent feed to the LIC was shut off and the 20 PSIG air purge for 10 seconds was insufficient to purge the agent line into the furnace. As the furnace cooled down, agent vapors were released into the chamber and pulled through the furnace system by the ID fan. Technical investigation conducted and procedures changed.
31-Oct-90	JACADS – (Unusual Occurrence)	Observation Corridor (09-123)	During feed rate capacity test the primary burner locked out on flamed failure. Possible agent migration through cracks in the wall which separates the LIC from the Observation Corridor. HVAC engineers performed smoke test and found leaks at the upper joint. Separation wall repaired.
15-Nov-90	JACADS – (Unusual Occurrence)	Perimeter Stations	No ACAMS agent alarms sounded in the "Red Hat" or any other JACADS area. No DAAMS confirmation was

Date	Facility	Location	Description
			received from any other operation. Investigation conclusion was no agent present. DAAMS tubes were contaminated.
30-Sep-91	JACADS – (Unusual Occurrence)	Lower Observation Corridor	Worker dropped a vial of dilute VX standard. Spill deconned and area monitored.
22-Dec-91	JACADS – (Chemical Event)	LSS Station # 77	Detection of VX in DAAMS samples taken from LSS station # 77. Contamination of the DAAMS valve, LSS-VALVE-868 silver fluoride pads assembly. Operator was wearing contaminated gloves and transferred contamination. LSS reconnected to a different manifold with zero readings. Operating procedures changed.
21-Jan-92	JACADS – (Unusual Occurrence) [PMCD]	DFS	Processing VX-filled M55 Rockets when a detonation occurred within the DFS, causing the kiln to stop rotating.
30-May-92	JACADS – (Unusual Occurrence)	Liquid	During slag removal and small sliver of glass punctured the operators Tyvex protective clothing cutting the operators calf. Blood draw was normal.
03-Jul-92	JACADS – (Unusual Occurrence)	Toxic Cubicle	Maintenance personnel enter toxic cubicle to replace an instrument and valve. When one of the agent transfer lines was cut, a dark oily substance leaked out. ACAMS readings began to rise. Personnel in level B left the toxic cubicle. Blood draw on entrants was negative. Operators re-entered in DPE suits and finished maintenance operation.
2-Jan-93	JACADS – (Unusual Occurrence) [PMCD]	Explosive Containment Room-A	During M60 105mm projectile processing within the ECR a fire occurred along the miscellaneous parts conveyor. Fire was contained within the ECR. Changes made to the equipment and increased frequency of ECR cleanup of residual explosives.
17-Mar-93	JACADS – (Chemical Event) [PMCD]	MDB	Raytheon Engineering and Constructors worker potentially exposure to Mustard Agent (HD). Worker developed blister(s) on leg after handling HD contaminated waste materials.
19-Mar-93	JACADS – (Unusual Occurrence)	MPB	120V AC extension cord over heated due to a short causing a small fire. The fire was extinguished with an ABC fire extinguisher
8-Dec-93	JACADS – (Chemical Event)	MDB	CON operator noted the agent level in ACS-tank-102 was not increasing. Draining process stopped and a remote camera scan of MPB conducted. Scan revealed agent discharging onto the floor. Investigation revealed a manual block value was closed on the ACS Tank. Agent backed up the piping and spilled onto the floor of the MPB. All agent maintained under engineering controls. Corrective actions taken to prevent reoccurrence
13-Dec-93	JACADS – (Chemical Event)	Lower Observation Corridor (LOC)	Entrants entrained agent into LOC via airlocks. Technical Investigation completed and operation procedures changed.
18-Dec-93	JACADS – (Unusual Occurrence)	SDS-Pump	Worker received a caustic burn (non-chemical agent) on his left hand. Protective clothing inadequate for task.

APPENDIX B 77

Date	Facility	Location	Description
07-Jan-94	JACADS – (Unusual Occurrence)	MDB - Exhaust Filter System	Momentary power failure and subsequent loss of ventilation in the MDB. No confirmed agent detected in Category C areas. Corrective action JACADS on own separate power distribution system.
13-Jan-94	JACADS – (Unusual Occurrence)	Lower Observation Corridor (LOC)	Entrants entrained agent into LOC via airlocks. ACAMS monitoring the LOC had a positive reading slightly above LOQ but below the TWA. Technical Investigation completed and operation procedures changed.
14-Mar-94	JACADS – (Chemical Event)	Explosive Containment Room B	Top feed gate did not open allowing cut rocket sections 2, 3, 4, and 5 to remain on top of the feed gate. Cut sections flashed causing a small fire. Deluge system water spray extinguished the fire. Rocket processing halted. All agent vapors maintained under engineering controls. Investigation conducted and corrective actions taken - install top feed gate brushes and program change to Logic Controller.
15-Mar-94	JACADS – (Unusual Occurrence)	MPF	During Capability test of Metal Parts Furnace (MPF) three- (3) pressure spikes were observed. Note all projectiles had the burster wells pulled.
23-Mar-94	JACADS – (Chemical Event) **[PMCD]**	Common Stack	LIC was being ramped down (controlled cooling operation) for slag removal. Minute amount of GB released via common stack. Technical Investigation completed and operation procedures changed.
20-Apr-94	JACADS – (Unusual Occurrence)	Lower Observation Corridor (LOC)	Entrants entrained agent into LOC via airlocks. ACAMS monitoring the LOC had a positive reading slightly above LOQ but below the TWA. Technical Investigation completed and operation procedures changed.
21-Apr-94	JACADS – (Unusual Occurrence)	LIC/MPF airlock	MDB filter unit taken off-line to perform maintenance. Air dampers on spare filter did not operate properly causing a fluctuation in the MDB. Agent maintained under engineering controls. Agent alarm in LIC/MPF airlock. Cause limit switch failure. Corrective actions taken.
19-Nov-94	JACADS – (Unusual Occurrence) **[PMCD]**	Explosive Containment Room	Detonation of rocket on fuzz shear causes agent migration to observation corridor. All agent vapor contained under engineering controls and exhausted through the MDB charcoal filter units.
2-Mar-95	JACADS – (Unusual Occurrence)	JACADS Perimeter	Unconfirmed perimeter DAAMS for GB.
17-Mar-95	JACADS – (Chemical Event)	Filter 404	Release of agent vapor from filter bank 404. Agent detected in temporary enclosure below the GPI level and registered within the range of 0.31 - 0.39 of the GPI level. Filter bank was off line and restarted and "air-washed".
17-Jan-95	JACADS – (Unusual Occurrence)	JACADS - MDB	Power failure with agent migration. No agent detected in Category C areas.
16-Aug-96	JACADS – (Unusual Occurrence)	Upper Observation Corridor	A low-level indication of GB vapor was detected in the Observation Corridors. Cause temporary loss of power.
28-Aug-96	JACADS – (Unusual Occurrence)	JACADS	Loss of power.

Date	Event	Location	Description
19-Dec-96	JACADS – (Unusual Occurrence)	Upper Observation Corridor	During a heavy rainstorm a ground fault occurred causing the power grid to become unstable. Loss of power with agent readings in an Observation Corridor.
22-Apr-97	JACADS – (Unusual Occurrence)	JACADS	CONR-110 rack 16 faulted resulting in loss of power.
17-Jul-97	JACADS – (Unusual Occurrence)	Filter 407	GB vapors in filter airlock for filter bank 407.
7-Sep-97	JACADS – (Unusual Occurrence)	MDB Observation Corridor	Indications of GB vapors in MDB Observation Corridors.
2-Feb-98	JACADS – (Unusual Occurrence)	Common PAS	ACAMS alarm at common PAS stack (due to promo w/ burster case intact being sent to MPF), no agent detected. ACAMS alarmed due to products of incomplete combustion.
16-Jan-00	JACADS – (Unusual Occurrence)	JACADS Treaty Trailer	Operational anomaly in JACADS Treaty Trailer. No agent operations at the time of reading. False positive.
04-Apr-00	JACADS – (Unusual Occurrence)	MDB	Removing contaminated clothes after deconned. Blood draws on workers indicated no depression of CHE.
11-Apr-00	JACADS – (Unusual Occurrence)	JACADS	Operational anomaly during training exercise.
23-Apr-00	USACAP 00-01 – (Chemical Event)	JACADS	Sample of liquid contained in a 55-gal drum had a positive reading for VX. Blood draws on 29 people indicated no depression of CHE. Technical investigation completed with recommended procedural changes.
19-Oct-00	USACAP 01-01 – (Chemical Event)	DFS PAS	Ash from DFS cyclone contained low levels of VX. 4 workers & 1 OPCW inspector sent for blood draw. All five individuals with no CHE depression.
05-Dec-00	USACAP 01-02 – (Chemical Event) **[PMCD]**	JACADS HDC Waste Bin	Agent detected in ash removed from the DFS. Agent containment spill pillows are the possible source of agent readings in ash. No personnel exposures.

Appendix C

List of Individual Incidents from the Chemical Weapons Working Group

Chemical Weapons Working Group
P.O. Box 467 Berea, KY 40403
606-986-7565 Fax: 606-986-2695
kefwilli@acs.eku.edu www.cwwg.org

Alabama
Burn Busters
Coosa River Basin Initiative
Environmental Justice Task Force
Families Concerned About Nerve Gas Incineration
Serving Alabama's Future Environment

Arkansas
Arkansas Fairness Council
Arkansas Public Policy Panel
Arkansas Sierra Club
Pine Bluff for Safe Disposal
Women's Action for New Directions

Colorado
Citizens for Safe Weapons Disposal
Sangre de Cristo Group of the Sierra Club

Indiana
Citizens Against Incinerating at Newport
Newport Study Group

Kentucky
Common Ground
Concerned Citizens of Madison County

Maryland
APG Superfund Citizens Coalition
Coalition for Safe Disposal
Concerned Citizens for Maryland's Environment

Oregon
GASP
Oregon Clearinghouse for Pollution Reduction
Oregon Sierra Club
Oregon Wildlife Federation

Utah
Families Against Incinerator Risk
Utah Sierra Club
West Desert Heal

Pacific
Pacific Asia Council of Indigenous Peoples
Pacific Friends Service Committee

Russia
Rainbow Keepers
Third Way Party
Center for Assistance to Environmental Initiatives
on Chemical Safety

(Partial List)

MEMO

To: The Chemical Events Committee of the National Research Council

From: Craig Williams on behalf of the Chemical Weapons Working Group

Re: Events at TOCDF involving suspected agent releases and worker/public/environmental exposures

Attached is a list of suspected agent events at TOCDF. By the advice of our counsel, we qualify this submission in the following way.

This is a managable list of events that should be investigated. It should not be construed to be an all-inclusive list of incidents that have occured or incidents that the CWWG either knows or suspects to have occured.

The incidents that are high-lighted in yellow are of particular concern to our coalition.

We are highly skeptical about PMCD's forthrightness in disclosing and accurately describing "chemical events" at TOCDF. Updates from PMCD on TOCDF operational incidents have decreased noticeably since the May 8-9, 2000 agent releases.

However, information from inside sources indicate they are experiencing continuing DFS feed gate problems, slag fires in the LIC and other events of unknown potential for agent releases. We request that the Committee review actual logs and CON operations reports for the past year and not rely solely on incident lists/reports provided by PMCD.

We also request that the Committee review actual logs and CON operations reports on the events listed in this attachment and not rely solely on incident reports provided by PMCD.

Also enclosed is a video of M-55 rocket processing, received from a source within TOCDF, which indicates significant amounts of agent volatization off the feed chute and kicker plate of the DFS. We request that the Committee look into the implications of what is recorded in this video to include, but not be limited to, the impact on the HVAC carbon filter design, DFS feed chute design and the potential for incomplete agent destruction due to inadequate residence time at temperature.

Thank you.

Craig Williams

Tooele Chemical Weapons Incinerator

Shutdowns /Incidents/Key Developments
Since Agent Operations Began August 22, 1996

- **August 24-26, 1996**-- Shutdown due to agent detection in the heating, ventilation, and air conditioning filter bank vestibules. Possible agent release into the environment.

- **September 9, 1996**-- Shutdown due to complete power failure in the plant. Possible agent release into the environment.

- **September 18, 1996**-- Shutdown due to potentially agent- contaminated decontamination fluid leaking through cracks in the concrete floor into electrical room below.

- **September 19, 1996**-- Shutdown due to Liquid Incinerator Slag Removal System malfunction during a shakedown trial burn.

  December 19, 1996-- Shutdown due to M-55 rockets jamming in the feed gates to the Deactivation Furnace.

- **January 20, 1997**-- Toxic spill in the 90-day storage yard improperly cleaned up. Haz-mat team called back to the plant prior to clean-up in "order to continue processing." Toxic material snow-plowed against the boundary fence and left.

- **January 26, 1997**-- Shutdown due to agent migration inside the observation corridors of the Munitions Disposal Building. Possible agent release to the environment.

- **February 6, 1997**-- Site masking alarm and/or stack alarm. Potential case of chemical warfare agent release or release of other related toxic chemicals (unidentified to date).

- **February 14, 1997**-- Site masking alarm and/or stack alarm. Potential case of chemical warfare agent release or release of other related toxic chemicals (unidentified to date).

- **March 13, 1997**-- Site masking alarm and/or stack alarm. Potential case of chemical warfare agent release or release of other related toxic chemicals (unidentified to date).

- **March 20, 1997**-- Army Project Manager, Tim Thomas admits in Federal Court to six confirmed nerve agent stack alarms. Nerve agent releases require shutdowns.

- **March, 24 1997**-- M-55 Rocket campaign halted due to trial burn failure for PCBs under Toxic Substances Control Act Requirements for 99.9999% DRE for PCBs. Public not notified until October 18, 1997.

- **April 10, 1997**-- Site masking alarm and/or stack alarm. Potential case of chemical warfare agent release or release of other related toxic chemicals (unidentified to date).

APPENDIX C

- **April 13-14, 1997--** Site masking alarm and/or stack alarm. Potential case of chemical warfare agent release or release of other related toxic chemicals (unidentified to date).

- **April 18, 1997--** Site masking alarm and/or stack alarm. Potential case of chemical warfare agent release or release of other related toxic chemicals (unidentified to date).

- **April 20, 1997--** Site masking alarm and/or stack alarm. Potential case of chemical warfare agent release or release of other related toxic chemicals (unidentified to date).

- **April 21, 1997--** During routine maintenance alarms sound indicating an unusually high agent reading (> 1200TWA) inside TOCDF.

- **April 22, 1997--** Site masking alarm and/or stack alarm. Potential case of chemical warfare agent release or release of other related toxic chemicals (unidentified to date).

- **April 23, 1997--** Shutdown due to "Notice of Insufficient Quality." Army Project Manager issues notice to curtail operations due to failure to follow operating procedures. Initially shutdown presented as "routine maintenance." The plant remains shutdown until June 15.

- **April, 25 1997--** Over 4,000 pages of official TOCDF documents arrive at CWWG office showing improper analysis, characterization, manifesting, tracking and disposal of Hazardous Waste leaving TOCDF.

- **May 3, 1997--** Site masking alarm and/or stack alarm. Potential case of chemical warfare agent release or release of other related chemicals (unidentified to date).

- **May 6, 1997--** Site masking alarm and/or stack alarm. Potential case of chemical warfare agent release or release of other related toxic chemicals (unidentified to date).

- **May 7, 1997--** Site masking alarm and/or stack alarm. Potential case of chemical warfare agent release or release of other related toxic chemicals (unidentified to date). Second occurrence on same day.

- **May 8, 1997--** Site masking alarm and/or stack alarm. Potential case of chemical warfare agent release or release of other related toxic chemicals (unidentified to date).

- **May 9, 1997--** Army admits to reporters that it "misled" the public about the cause of the six week shutdown.

- **May 13, 1997--** Site masking alarm and/or stack alarm. Potential case of chemical warfare agent release or release of other related toxic chemicals (unidentified to date).

- **May 14, 1997--** Site masking alarm and/or stack alarm. Potential case of chemical warfare agent release or release of other related toxic chemicals (unidentified to date).

- **May 15, 1997--** Accusations of illegal burning of Lewisite made by CWWG. Army first denies illegal burning then later admits to having burned some containers that previously contained Lewisite. Plaintiffs' evidence indicates Lewisite was burned.

- **May 24, 1997--** Site masking alarm and/or stack alarm. Potential case of chemical warfare agent release or release of other related toxic chemicals (unidentified to date).

- **May 28, 1997--** Citizens from Oregon and Kentucky are escorted through TOCDF into an area with a GB-contaminated bomb casing present. Citizens are not notified until after an anonymous call from TOCDF notifies Utah DEQ and an OR citizen.

- **July 6, 1997--** Shutdown due to Pollution Abatement System (PAS) blockage. Amount of agent and other toxics emitted unknown.

- **August 1, 1997--** Former Chief Safety Officer, Steve Jones is ruled for in his Dept. of Labor "Wrongful Termination Action." Judge awards Jones his job back and $500,000 or no rehiring and $1 million. Judge calls EG&G managers "liars."

- **August, 1997--** Site masking alarm and/or stack alarm. Potential case of chemical warfare agent release or release of other related toxic chemicals (unidentified to date).

- **September 8, 1997--** Site masking alarm and/or stack alarm. Potential case of chemical warfare agent release or release of other related toxic chemicals (unidentified to date).

- **September 9, 1997--** Former Chief of Hazardous Waste Management, Trina Allen wins on her discrimination part of Dept. of Labor claim against EG&G. Allen is awarded $5,000. A Hearing on the merits of the remainder of her claims is scheduled for December.

- **September 12, 1997--** Army admits, in documents sent to Utah DEQ, that it has been burning Lewisite (L) contained in Ton Containers of GB previously contaminated with "L," confirming allegations made by citizen activists that TOCDF has illegally operated.

- **September 14, 1997--** Site masking alarm and/or stack alarm. Potential case of chemical warfare agent release or release of other related toxic chemicals (unidentified to date).

- **September 18, 1997--** Site masking alarm and/or stack alarm. Potential case of chemical warfare agent release or release of other related toxic chemicals (unidentified to date).

- **September 30, 1997--** Site masking alarm and/or stack alarm. Potential case of chemical warfare agent release or release of other related toxic chemicals (unidentified to date).

- **October 1, 1997--** Site masking alarm and/or stack alarm. Potential case of chemical warfare agent release or release of other related toxic chemicals (unidentified to date).

APPENDIX C

- **October 2, 1997**-- Site masking alarm and/or stack alarm. Potential case of chemical warfare agent release or release of other related toxic chemicals (unidentified to date).

- **October 6, 1997**-- Site masking alarm and/or stack alarm. Potential case of chemical warfare agent release or release of other related toxic chemicals (unidentified to date). Second and third occurrences on the same day.

- **October 11, 1997**-- Site masking alarm and/or stack alarm. Potential case of chemical warfare agent release or release of other related toxic chemicals (unidentified to date).

- **October 12, 1997**-- Site masking alarm and/or stack alarm. Potential case of chemical warfare agent release or release of other related toxic chemicals (unidentified to date).

- **October 16, 1997**-- Site masking alarm and/or stack alarm. Potential case of chemical warfare agent release or release of other related toxic chemicals (unidentified to date).

- **October 17, 1997**-- Site masking alarm and/or stack alarm. Potential case of chemical warfare agent release or release of other related toxic chemicals (unidentified to date).

- **October/November 1997**-- Sources inside TOCDF (who wish to remain anonymous) communicate to CWWG several shutdowns/incidents at TOCDF due to computer malfunctions, slag build-up in the PAS, numerous agent migrations within the facility, and alarm ring-offs in the common stack, MDB and HVAC stack (averaging 2-3 per week).

- **November 2, 1997**-- Site masking alarm and/or stack alarm. Potential case of chemical warfare agent release or release of other related toxic chemicals (unidentified to date). Second occurrence on the same day.

- **November 6, 1997**-- House Government Reform and Oversight Committee unanimously approves Human Resources Subcommittee Report on Gulf War that concludes exposures to low-level chemical agents (lower than the amount set as "acceptable" at TOCDF) caused or contributed to Gulf War Illness.

- **November 18, 1997**-- TOCDF is cited for 25 violations by the Utah Department of Environmental Quality. Citations included "numerous instances of noncompliance," but not enough to shut them down, according to a DEQ spokesperson.

- **November 26, 1997**-- Site masking alarm and/or stack alarm. Potential case of chemical warfare agent release or release of other related toxic chemicals (unidentified to date).

- **November 27, 1997**-- Sources inside TOCDF (who wish to remain anonymous) communicate to CWWG that both Liquid Incinerators (LICs) are "down" due to malfunctions. According to sources, one of the LICs has been down for over a month.

Chronic problems with the Brine Reduction Area (BRA) and the Pollution Abatement System (PAS) continue to plague TOCDF.

- **November 30, 1997**-- Site masking alarm and/or stack alarm. Potential case of chemical warfare agent release or release of other related toxic chemicals (unidentified to date).

- **December 1, 1997**-- Site masking alarm and/or stack alarm. Potential case of chemical warfare agent release or release of other related toxic chemicals (unidentified to date).

- **December 2, 1997**-- Site masking alarm and/or stack alarm. Potential case of chemical warfare agent release or release of other related toxic chemicals (unidentified to date).

- **December 3, 1997**-- Site masking alarm and/or stack alarm. Potential case of chemical warfare agent release or release of other related toxic chemicals (unidentified to date). Second and third occurrences on the same day.

- **December 5, 1997**-- Site masking alarm and/or stack alarm. Potential case of chemical warfare agent release or release of other related toxic chemicals (unidentified to date).

- **December 7, 1997**-- Site masking alarm and/or stack alarm. Potential case of chemical warfare agent release or release of other related toxic chemicals (unidentified to date).

- **December 20, 1997**-- Site masking alarm and/or stack alarm. Potential case of chemical warfare agent release or release of other related toxic chemicals (unidentified to date).

- **December 21, 1997**-- Site masking alarm and/or stack alarm. Potential case of chemical warfare agent release or release of other related toxic chemicals (unidentified to date).

- **December 26, 1997**-- Site masking alarm and/or stack alarm. Potential case of chemical warfare agent release or release of other related toxic chemicals (unidentified to date).

- **January 1, 1998**-- After almost 20 years of denial, a document surfaces showing that the Army had proof as early as 1970 that the 1968 sheep kill in Skull Valley was a direct result of nerve agent exposure by the Army. Recent depositions in CWWG federal lawsuit disclose that Army officials have come to the conclusion that the sheep were killed as a result of the combined effect of the nerve agent sprayed and pesticides already present.

- **January 28, 1998**-- Site masking alarm and/or stack alarm. Potential case of chemical warfare agent release or release of other related toxic chemicals (unidentified to date).

APPENDIX C

- **January 31, 1998**-- Site masking alarm and/or stack alarm. Potential case of chemical warfare agent release or release of other related toxic chemicals (unidentified to date).

- **January 31, 1998**-- Department of Labor Administrative Law Judge, Samuel J. Smith orders EG&G to reinstate whistleblower Trina Allen and to "cease and desist" any retaliation against her and other employees for protected activities in the conduct of performance of their duties.

- **February 1, 1998**-- Site masking alarm and/or stack alarm. Potential case of chemical warfare agent release or release of other related toxic chemicals (unidentified to date).

- **February, 1998**-- Worker on the hazardous waste crew discloses that vials filled with chemical warfare agent, and having contamination on the outside of the vials, were misplaced for a few days. Later they were found in the toxic maintenance area. Had these vials not been located, they would have been sent off-site as generic waste to be disposed of at a commercial facility.

- **February 4, 1998**-- Site masking alarm and/or stack alarm. Potential case of chemical warfare agent release or release of other related toxic chemicals (unidentified to date).

- **February 12, 1998**-- According to the Deseret News, TOCDF experienced an "all mask" alarm situation while attempting to go back on line after a 30-day "routine maintenance" period. According to Utah DEQ, "They (TOCDF) did confirm that this was not a false alarm."

- **March 11, 1998**-- Five TOCDF employees fall ill with symptoms of dizziness, headache and nausea. Army officials say industrial materials are suspected.

- **March 16, 1998**-- Sources inside TOCDF (who wish to remain anonymous) communicate to CWWG that alarms have been sounding regularly in the Unpack Area during the recently initiated MC-1 bomb campaign. These sources also claim that workers in the area are not wearing "Level B" protective gear as required.

- **March 30, 1998**-- Shutdown occurs when the metal parts furnace (MPF) overheats due to feeding an illegal amount of nerve agent GB into the furnace. The ACAMS alarm in the MPF duct rings off at approximately 850 times the allowable stack concentration for agent. The ACAMS alarms in the common stack register a large chemical spike. No DAAMS tubes are located at the duct ACAMS to confirm for agent and no one knows if the DAAMS tubes in the stack at the time of the incident have been analyzed to confirm for agent. Army officials claim that agent did not go out the stack, but can't prove that the large amount of chemical released was not agent. The chemical plume was neither quantified or qualified.

- **November 21, 1998 - January 7, 1999**-- There are seven instances of Unpack Area ACAMS alarms with individuals wearing inadequate protective clothing.

- **November 25, 1998**-- Vapor leaks of GB (sarin) are detected from three 105mm projectiles. The agent is detected while one of the projectiles is being processed into an incinerator.

- **November 28, 1998**-- Another vapor leak of GB is detected from a 105mm projectile which is in an on-site container.

- **December 4, 1998**-- It is reported that 24 vapor leaks have been detected in the past two months, all involving 105mm projectiles. Each of these 16-inch long bullet-shaped objects contains .17 gallons of GB. 16 of the leaks were detected after the projectiles were transferred from the storage igloos to the incinerator building. Eight of the leaks occurred when crews were removing a heavy bolt screwed into the nose of the projectile.

- **December 13, 1998**-- Liquid Incinerator is shut down after 140 gallons of GB (sarin) are spilled while being fed into the incinerator, raising serious questions about the engineering and design of the technology.

- **April 13, 1999**-- Shutdown occurs when TOCDF back-up power system fails after Depot-wide outage. This failure compromised the negative air flow system, fans leading to the stacks and other critical systems. Possible agent release to the environment and worker exposure.

- **April 16, 1999**-- There is confirmed agent reading in the DFS Cyclone Enclosure which is adjacent to the outside--possible agent release to environment.

- **May 1, 1999**-- ACAMS alarms at 508.4 twa in Unpack Area with three workers in inadequate protective clothing. During feed stop on LIC 1, there is ACAMS duct alarm at 1.26 asc and a stack alarm at .34 asc.

- **May 5, 1999**-- Agent vapor leak forces workers out of certain areas within TOCDF.

- **May 21, 1999**-- Agent migrates from a Level A to a Level C area where agent is not supposed to be present. The ACAMS reading in the Level C area is 75 times the alarm point of .2twa. After alarm for agent presence, seven workers have to don the masks that are at their hips and evacuate. They are not in adequate protective clothing. Army officials testified in federal court that they don't know if any agent escaped to the outside environment during this incident.

- **May 24, 1999**-- Workers removing nose closures from 105mm projectiles encounter liquid agent in a burster well where liquid agent isn't anticipated. Workers are not in adequate protective clothing. According to testimony of Project Manager Tim Thomas the ACAMS rang off at approximately 1900twa--50 times the maximum level of agent for the clothing the workers were wearing.

- **May 26, 1999**-- Workers in the Toxic Maintenance Area are removing plastic bags of waste when the ACAMS alarms at 1985twa causing them to evacuate. One of the bags containing liquid agent is ripped. Again workers are not in adequate protective clothing. Workers still ring off positive for agent after doffing their clothing in the airlock. They then ring off positive after being rinsed with water and still ring off positive after a further rinsing with bleach. They have final positive readings as they depart from the airlock.

- **June 4, 1999**-- County-wide power outage causes TOCDF negative air flow system (HVAC) to go down. It is 25 minutes before the emergency backup power system comes on. Backup power is supposed to come on automatically. Loss of the HVAC system causes agent to migrate into Level C areas where agent isn't supposed to be. There are 3 site masking alarms during the power outage event. Army officials testified in court that they don't know if agent migrated to the outside environment. (See attached Guello/Burton memo.)

- **June 5-June14, 1999**-- TOCDF is in "Stand Down" by order of Chem Demil according to testimony of Col. Joseph Huber in federal court. No munitions are processed during this period while a Review Team from Aberdeen is looking at recent agent events at TOCDF.

- **June 14, 1999**-- TOCDF starts up after "Stand Down." Processing of M-55 rockets is resumed. However TOCDF is shut down again because within 6 hours of start-up, allowable feed rate for rocket processing is violated.

- **August 1 - September 13, 1999**-- There are 19 "potential" worker exposures. 4 are in Level B clothing. 11 are due to rips in protective ensembles or gloves and 4 workers are present in over 500 IDLH atmosphere.

- **August 9, 1999**-- Tangled air hoses prevent DPE entrants from reaching egress air locks.

- **August 9, 1999**-- Worker exposed to nerve agent with tear in protective suit not seen at clinic until three and a half hours later.

- **August 25, 1999**-- ACAMS in Unpack Area alarms at .21 twa. The ACAMS heat trace is discovered to be burning.

- **Week of Aug. 31-Sept. 4**-- DFS feed chute gets jammed with rocket pieces. Site team shuts down DFS to change out warped section of feed chute. Angle irons used to dislodge previous jam get jammed in chute also.

- **Week of Aug. 31-Sept. 4**-- Internal report blames cracks in concrete floor of MDB for decon seepage into electrical room and states that new cracks continue to be identified.

- **September 9, 1999**-- Cleanliness and organization of toxic areas is so bad that processing is shut down for 59 hours to get housekeeping issues straightened out.

- **September 20, 1999**-- Internal report states that SOPs are happening too quckly for people to keep up and more often than not, new SOPs are not being carried out.

- **September 21, 1999**-- Internal report reveals that workers are performing unapproved SOP of hitting wooden pallets with a steel mallet to loosen pallet covers which results in projectiles falling from pallets onto to UPA floor.

- **September 23, 1999**-- Internal report blames poor contamination control and inadequate decontamination attempts for high levels of contamination in airlocks.

- **September 27, 1999**-- Control room operator discovers that two "pressurized" ton containers have been in the 90-day storage site for greater than the allowed time. Report of incident states that information transmitted to management after discovery of tons is "less than adequate,...training received on environmental inspections is inadequate." Incident results in a Government Nonconformance Report and an EG&G Deficiency Report.

- **November 3, 1999**-- Manager orders worker to make DPE entry against advise of monitoring team who informed him of questionable ACAMS reliability in hot area.

- **November 3, 1999**-- TOCDF system engineer calls LIC slag removal system "fatally flawed"--engineers have to "jumper" the system code to get the system to operate correctly.

- **November 3, 1999**-- ACAMS alarm of .37 in LIC Secondary room goes unnoticed for 2 hours during which time several workers enter the room.

- **November 9, 1999**-- ACAMS in EHM alarms at 1.23 twa with unmasked workers present. Personnel are told to exit into airlock but are not told seriousness of situation and are not told to mask.

- **December 6, 1999**-- The protective suits of two workers are melted during slag removal operation in one of the liquid incinerators.

- **December 7, 1999**-- In a fire in the upper gate of the deactivation furnace feed chute, three rocket sections burn. Flames are also seen on the floor and at the shear blade. The time of the fire is uncertain "due to unreliability of the fire sensor." Instructions have been given to avoid leaving rocket sections on the upper gate "even if it means burning them in the chute." Three hours earlier, the lower gate malfunctioned and resulted in a stop feed. It takes ten days to prepare report on the incident.

- **February 20, 2000**-- Two workers exposed to nerve agent GB when it leaks into room where they are working.

- **February 23, 2000**-- 40 to 45 gallons of molten slag spills from a drum and starts a fire that burns the covering of the concrete floor and electrical equipment in a secondary room of the liquid incinerator.

- **April 30, 2000**-- A maintenance man just happens to walk past the Cyclone Ash Bin Enclosure of the Deactivation Furnace (DFS) and notices smoke, heat and a bulged out door. There is a fire going on that no one had detected. The fire ignites and decomposes the charcoal in the filter system of the Ash Bin and would have entered the filter banks of the MDB if it hadn't been discovered by a worker out for a walk. The fire was precipitated when the blind flange was installed in preparation for an entry into the DFS to clear a jam in the Heated Discharge Conveyor.

- **May 8, 2000**-- After workers finish maintenance on deactivation furnace feed chute there is confirmed release of nerve agent GB out of common stack into environment at 11:26 pm. Army reports that afterburner was blown out due to malfunction of air flow meter which was clogged with liquid. Stack alarm rings off for about 20 minutes at somewhere between 3.7 and 8.6 times the allowable stack concentration of GB. Army reports on the alarm reading are inconsistent. Mysteriously, after stack alarm rings off, alarm in furnace duct leading to stack rings off for agent presence. Facility managers say they have no clue as to why sequence of alarms was apparently backwards.

  May 9, 2000-- A confirmed release of GB to the environment takes place at about 1:15 am (less than an hour and a half after the confirmed release May 8) when workers attempt to relight deactivation furnace afterburner. Local emergency officials not notified until four hours after first GB release. Decision made by Army manager not to immediately notify local officials is in violation of Army SOPs, facility's operating permit granted by State of Utah and agreements with local emergency responders. Facility is shut down until investigation team, headed by the Army, makes final report on the incidents. EG&G manager predicts investigation will lead to physical modification, not just new SOPs. Shut down could last several weeks.

- **June 6, 2000**-- TOCDF is still shut down. It is reported that the facility's shut down is costing about $285,000 per day--totaling almost $8 million to date.

- **July 26, 2000**-- TOCDF has been shut down for 79 days. At $285,000 per day--the cost so far is more than $22 million.

- **July 28, 2000**-- The Utah DEQ authorizes the restart of the two liquid incinerators and the metal parts furnace after the entire facility had been shut down for 81 days following the May 8-9 agent releases.

- **September 19, 2000**-- The Utah DEQ authorizes the restart of the DFS after it had been shut down for 133 days following the May 8-9 agent releases.

- **October 19, 2000**-- At the Utah CAC meeting, it is stated that there had been 97 agent alarms at TOCDF since May 8. 14 of the alarms were in the common stack.

- **November 16, 2000**-- At the Utah CAC meeting, it is stated that there had been 41 agent alarms at TOCDF since October 19. Three of the alarms were in the common stack and five were in ducts leading to the common stack.

- **November 25, 2000**-- The nerve agent GB (sarin) is detected in employees' work clothes. The workers come in from inspecting filters outside in cold weather and apparently the sarin begins vaporizing as their clothing warmed up.

- **December 9, 2000**--Agent break through in HVAC filter bank. ACAMS readings of 3.01.

Appendix D

List of Individual Incidents from Calhoun County Commission, Anniston, Alabama

CALHOUN COUNTY COMMISSION
1702 NOBLE STREET, SUITE 103
ANNISTON, ALABAMA 36201
TELEPHONE (256) 241-2800
FAX (256) 231-1744

COMMISSIONERS

JAMES A. DUNN
District 1
ROBERT W. DOWNING
District 2
JAMES ELI HENDERSON
District 3
RANDY WOOD
District 4
LEA FITE
District 5

KENNETH L. JOINER
Administrator/Treasurer

THOMAS M. SOWA
County Attorney

TO: National Research Council Committee on Chemical Events

FROM: Calhoun County Commission

RE: <u>Issues concerning the Scope and Statement of Task (SOT) of the "Committee on Evaluation of Chemical Events at Army Chemical Agent Disposal Facilities."</u>

DATE: June 21, 2001

The Calhoun County Commission respectfully requests that the Committee examine and report on the following issues:

I   <u>PROCESS TECHNOLOGY</u>

(A)   <u>Modified Baseline:</u> The Commission requests that the Committee assess the validity of deploying the "Modified Baseline" technology, as proposed in Pueblo, Colorado, to the Anniston facility as well as the Tooele, Pine Bluff and Umatilla sites, as referenced in Attachment 1 of the Operations Schedule Task Force 2000 – Final Report (October 20, 2000 §'s 1, 2, 7, 10 and 13).

In the case of "Modified Baseline," it is our understanding that projectiles, with the explosives removed, will be processed and that this will result in significantly larger quantities of agent being treated than the "baseline" technology was designed to accommodate. We understand that little information is available on the impact to the downstream pollution abatement system's capability to handle the larger volumes of off-gases, temperature spikes, as well as residence times for agent destruction. Also, it is our understanding this approach creates significantly greater quantities of agent within the system than is contemplated within the "baseline" design, leading to the potential for higher amounts of agent to be released during upset conditions, operational malfunctions, and raising serious questions surrounding steady state operational capabilities, given the delicate air flow balance required between the HVAC and combustion systems. To the best of the Commission's knowledge, none of these factors has been addressed in a Safety or Hazard Analysis or in the Health Risk Assessments for ANCDF.

(B) Gelled Rockets: The Commission requests that the Committee assess the validity of the Permit Modification currently under consideration for ANCDF which would allow trial burn processing rates of between 30 to 34 M-55 gelled rockets per hour with no agent draining or explosive reconfiguration prior to processing (i.e.: fully agent loaded, explosive and propellant contained chopped rockets fed to the deactivation furnace.) Under this approach the "unknowns" appear to be even greater in number and potentially more severe from a safety perspective than the "Modified Baseline" concept. In a 1991 report on the Cryofracture method of destruction, the National Research Council state, "Unsteady, very rapid burning of explosives and propellant elements would lead to a variable residence times for agents in downstream components, thus making complete combustion difficult to achieve. The combustion of so many different types of components simultaneously, with the potential for generating undesirable complex gases or solids in the process, plus the strong corrosive nature of the chemical agents make the use of a common kiln a most questionable procedure from the standpoint of both efficiency and safety." (emphasis added). Demilitarization of Chemical Weapons by Cryofracture: A Technical Assessment; National Research Council; 1991. The modifications being proposed for ANCDF create the exact set of circumstances which were of such deep concern in the NRC Cryofracture report; A single kiln would be processing agent, explosives, propellant and metal parts simultaneously.

These are examples of the types of issues under the "Process Technology" portion of the SOT the Commission requests be addressed by the Committee. A more extensive list may be provided within the two month period of time allotted by the Committee.

II. DESIGN CHANGES:

(A) Isolation Valves (Knife Gates): The Commission would like the Committee to include a recommendation on whether the inclusion of isolation valves at ANCDF is an appropriate modification of the incinerator from a safety and operational perspective. Subsequent to the release of agent GB at Tooele on May 8 and May 9, 2000, PMCD implemented the incorporation of a Control Room operated isolation and air (bleed) intake valve in the ducting between the DFS kiln and the Afterburner (AFB) to prevent unburned agent from escaping to the Pollution Abatement System and then out the common stack. (Approval for Restart of the Deactivation Furnace Letter from Utah DEQ to Commander, Deseret Chemical Depot; Item EG&G #12/Army #5; page 3; September 18, 2000). According to the above referenced document, "the valve has been function tested and is operational."

A similar design modification has been incorporated into the ANCDF. The Commission requests the Committee to consider the process flow implications of this modification of the incinerator in regard to the potential backpressure which could occur should the valve have to be deployed; as well as the characterization, route and final disposition of the air that would be released through the "air bleed" action and other associated impacts on process flow as a result of this design change. Specifically, the Commission would like to know the answer to the following questions: What kind of backpressure will be created when the operator drops the gate? What happens? It would appear to the Commission that the air has to go somewhere.

(B) <u>Safety or Hazard Analysis:</u> The Commission also would like to know if a Safety or Hazard Analysis has been done for this isolation valve design change and whether there are any Health Risk Assessment implications concerning this particular modification.

These are two examples of the types of issues under the "Design Change" portion of the SOT the Commission requests be addressed by the Commission. A more extensive list of additional design change issues may be provided within the two month period of time provided by the Committee.

III MANAGEMENT

(A) The Commission requests the Committee include within its management review the issues raised and conclusions reached in the following reports: all GAO Reports between 1990 – 2000; Overarching Issue Assessment Annual Report for FY 1998 (December 7, 1998); Army I.G. Report – "Command and Control Structure at Current and Future Army Chemical Storage and Demilitarization Sites (March 23, 1998); Report submitted to Congress by Undersecretary of Defense Gansler on behalf of Secretary of Defense Cohen (Report of the Secretary of Defense to the Congressional Defense Committees on the Management of the Chemical Demilitarization Program, April 2000).

These are examples of the types of documents and the issues under the "Management" portion of the SOT the Commission requests be considered by the Committee. A more extensive list may be provided within the two month period of time provided by the Committee.

IV <u>TOXICITY RISK MANAGEMENT AND SAFETY PROGRAMS</u>

(A) <u>Acute Agent Toxicity:</u> The Commission requests the Committee include within its Risk Management and Safety Review the latest information available on acute agent toxicity available from all of the NRC Committees which have examined this issue. In particular the Commission requests the Committee to focus on the acute agent toxicity information included in the NRC report, "Review of Acute Human-Toxicity Estimates for Selected Chemical-Warfare Agents" (National Academy Press; 1997) and the Army's CDEPAT Report, "Review of Existing Toxicity Data and Human Estimates for Selected Chemical Agents and Recommended Human Toxicity Estimates for Selected Chemical Agents and Recommended Human Toxicity Estimates Appropriate for Defending the Soldier" (1994). Additionally, the Commission asks the Committee to incorporate into this section a Review and Finding on the Chemical Warfare Agent sections of EPA's proposed "Acute Exposure Guideline Levels for Hazardous Substances; Proposed AEGL

Values" (Federal Register/Vol.66, No.85/Wednesday, May 2, 2001; pages 21940 – 21964).

(B) <u>Low Level Exposure of Agents:</u> The Commission also requests the Committee include within its Risk Management and Safety Review the incorporation of the latest information available on the impact of low-level exposure of agents. The Commission requests this review include, but not be limited to, the issues raised in "DoD Strategy to Address Low-Level Exposures to Chemical Warfare Agents" (May 1999).

(C) <u>Hazardous Waste Combustion:</u> The Commission requests the Committee include a review of the 1999 National Research Council report, <u>Waste Incineration and Public Health</u>, which made an assessment of the relationship between hazardous waste combustion and human health; the design, citing and operating conditions of combustion facilities; an overview of human exposure to pollutants the risk assessments and other methodologies used to measure potential exposures. The report notes that "Some of the available assessments may now be considered inadequate for a complete characterization of risk, for example, due to their failure to account for changes in emissions during process upsets, or because of gaps in and limitations of the data or techniques of risk assessment available at the time."

(D) <u>Hazardous Waste Emissions:</u> The Commission requests the Committee include a review of EPA's June 1998 document, <u>Development of a Hazardous Waste Incinerator Target Analyte List of Products of Incomplete Combustion</u>, which concluded that current sampling methods for characterizing hazardous waste incinerator emissions "provide an incomplete picture of the emission profile," and that a large number of products of incomplete combustion (PICs) remain unidentified; therefore the health effects of these compounds are unknown.

These are examples of the types of documents and the issues under the "Risk Management and Safety Review" portion of the SOT the Commission requests be considered by the Committee. A more extensive list may be provided within the two month period of time provided by the Committee.

V <u>IMPROVEMENTS TO OPERATIONAL ACTIVITIES</u>

(A) <u>ACAMS/DAAMS:</u> The Commission requests the Committee investigate the inadequacies of the ACAMS/DAAMS monitoring system with a particular emphasis on the DAAMS analysis agent confirmation methodology deployed by PMCD.

(B) <u>Real Time Monitoring:</u> The Commission requests the Committee consider the fact that Continuous Emissions Monitoring System (CEMS technology) exists in the method known as "EPA Method TO-16: Long-Path Open-Path Fourier Transform Infrared Monitoring of Atmospheric Gases" and other advanced real-time monitoring systems. (i.e., real time online MS/MS systems). This technology is a critical component of a proper safety regimen, and we request that you review and made a recommendation on the applicability of this technology for ANCA.

These are examples of the types of issues under the "Improvements to Operational Activities" portion of the SOT the Commission requests be considered by the Committee. A more extensive list may be provided within the two month period of time provided by the Committee.

VI   INPUT BY CITIZEN GROUPS

   (A)   Chemical Events:

   (1)   Identifying Events to be Reviewed: There is a high degree of skepticism in communities where chemical stockpiles are located that many "incidents" are not being disclosed to the public and that the decision by PMCD on whether or not to have these incidents formally characterized as "chemical events" is often based on the potential negative impact on the program's image rather than the actual circumstances surrounding the incidences. The Commission therefore requests that the Committee insure the lists of "incidents" that have complied by these citizen groups be compared with the list of "Chemical Events" provided to the Committee by PMCD.

   (2)   Agent Releases: In addition to the broader definition of a "Chemical Event" (wherein agent escapes "engineering controls") the Commission also requests an in-depth review of actual releases from JACADS and TOCDF that have been documented by citizens groups, as well as those identified as potential releases or potential releases by citizens groups but denied by PMCD. A few examples of these potential releases:

   (a)   August 26, 1996:   TOCDF initiated operations and the facility was shut down within 24 hours due to what citizens believe was an agent release from the HVAC filter area;

   (b)   October/November 1997:   Information concerning major computer problems at TOCDF and related incidents was communicated to citizen groups. No incident reports were released to the public although numerous common stack alarms were said to have gone off during the period.

   (c)   March 30, 1998 Incident: On July 23, 1999 the NRC Stockpile Committee was supplied with more than 1000 pages of data on the March 30, 1998 incident at TOCDF with the assurance that this incident would be investigated and reported on. To date no such investigation or report has been made public.

   (d)   June 4, 1999 Incident: A global power outage occurred at TOCDF. The automatic back-up power system failed and was not brought on line until 25 minutes after power outage. Documents indicate negative pressure capability during power loss can only be maintained for approximately 10-12 minutes; however, TOCDF was without power for 25 minutes. A memo in response to a Senate inquiry of the incident states that power was restored in 12 minutes, but the internal PMCD document written immediately after the incident states the facility was without power for 25 minutes.

(e) <u>December 9, 2000</u>: A HVAC filter bank experienced agent breakthrough at TOCDF. The ACAMS reading was 3.01 TWA.

(f) <u>2001:</u> Public information on TOCDF operations has decreased noticeably since the September 2000 House Armed Services Procurement Subcommittee Hearings on the May 8th, 2000, agent release at TOCDF. Information from sources at TOCDF indicate continued DFS feed gate problems, several slag fires in the LIC and other possible incidents of potential agent releases. The Commission requests that the Committee review actual logs and CON operations reports and <u>not</u> rely strictly on incident lists provided by PMCD in determining whether or not there is any validity to these allegations.

(B) <u>Exposure and Medical Treatment for Workers:</u> Citizen groups remain concerned about the impacts of low-level and other levels of exposure to workers at the CDF's, although the PMCD continues to publicly state that "no worker has ever been exposed to agent." The Commission shares the concerns of these citizen groups.

A few examples of the low level and other levels of exposure to workers at the CDF's which the Commission respectfully requests the Committee examine and report on are:

(1) <u>November 1996:</u> A NREMT/Health Technician at TOCDF has alleged to have been harassed by plant supervisors for suggesting suspected organophosphate poisoning of TOCDF workers based on the technician having observed increased numbers of bradycardias (slow heart rates) and atrioventricular blocks (delays or block in cardiac electrical conduction) amongst these TOCDF workers.

(2) <u>April 1997:</u> Two workers at TOCDF apparently were exposed to agent while in the TMA. After being decontaminated and stripped down to their underwear, ACAMS alarms still rang off above non-zero levels. Exact ACAMS levels remain unknown.

(3) <u>April 1997:</u> A JACADS Clinical Lab Technician sent a letter to the U.S. Senate alleging worker safeguards at the facility were being ignored and covered up by contractors.

(4) <u>May 26, 1999</u>: TOCDF – Occurrence Report No: 99-05-26-A1: After repeated decontamination and stripping down to bare skin of four workers ACAMS alarms continued to ring off in the "B" airlock.

(5) <u>April 11, 2000:</u> A RDC Investigation Report prepared April 18, 2000 stated that, during an emergency exercise, medical personnel on JACADS were

potentially exposed to agent but were not decontaminated and, according to affidavits, told not to report the incident.

The Commission requests these incidents be closely examined and investigated because recent scientific evidence indicates low-level non-acute exposures to nerve agents can cause long term health problems. There is also evidence which strongly suggests that blood sampling (the method used by PMCD to determine worker exposure), may not be an inadequate method for confirming whether or not workers have been exposed to agent.

(C) <u>Army Audit Agency (AAA) Report:</u> The Commission requests the Committee review the more than 3000 design changes noted in the AAA Report with particular attention to the lack of assessment of the effect of these design changes on operations. Modifying components within the overall design of the plants without consideration of their impacts on up or down stream functional capability would appear to contradict sound management policy and could be potentially dangerous from an operational standpoint.

According to the "Production Milestones" section of the SOT, a "Report Outline" is already done and a "Concept Draft" is due in just 14 days. The Committee has assured Congressman Riley and his staff that Alabama citizens will be involved in the front and back ends of the Committee's efforts, as in inputting into the scope and issues they would like the Committee to investigate. How is it that the Outline is already done and the Draft will be completed in 2 weeks, particularly in light of this meeting being the first opportunity for the Commission and local citizens to present matters to the Committee?

Appendix E

Additional Information Concerning Risk

QUANTITATIVE RISK ASSESSMENT

The Army's quantitative risk assessment (QRA) for the Tooele Chemical Agent Disposal Facility (TOCDF) estimates the risk to the public and to workers from accidental releases of chemical agent associated with all activities during storage and throughout the disposal process (U.S. Army, 1996a). Activities associated with the disposal process include:

- munitions handling in preparation for transport to the disposal facility
- transport of munitions to the disposal facility
- the disposal processes

The QRA includes all identified potential causes of release, except for intentional acts, such as sabotage and terrorism. Releases resulting from both internal initiating events (events that originate inside the facility or that result directly from activities during the disposal process) and external events (such as earthquakes, aircraft crashes, and tornadoes) are included.

The factors in developing a QRA for a chemical agent and munitions storage and destruction site are shown in Figure E-1, which shows as the two primary sources of risk (1) the stockpile itself (storage risk) and (2) the destruction of the stockpile (processing risk). The actual risk posed by either or both sources depends on whether or not risk-initiating events occur.

Stockpile-related risks from sabotage, terrorism, and war are reportedly evaluated and managed by specific government agencies and are not considered in publicly available site-specific risk assessments.

Storage Risk

The stockpile is hazardous principally because of the inherent toxicity of the anti-cholinesterase nerve agents, GB and VX, and mustard agents, H, HD, and HT. Because of its toxicity and volatility, agent GB presents the greatest hazard offsite. The risks associated with stockpile storage are almost all related to releases of agent as a result of either internal events—such as handling accidents during stockpile manipulation and maintenance, the deterioration of containment systems, the spontaneous detonation of munitions, or the spontaneous ignition of propellant—or external events such as natural disasters and airplane crashes.

Processing Risk

Agent destruction imposes risks above and beyond the inherent risks associated with the existence and maintenance of the chemical agent and munitions stockpile. The transportation of chemical agent from storage to the destruction facility, the unpacking and disassembly of munitions and containment systems, and the actual process of agent destruction are activities during which agent could potentially be released. Like the storage risk, the predominant processing risk is associated with agent toxicity, although the quantities of agent being processed at any given time are small compared with the original inventories in the stockpile and are much better protected than in the stockpile.

Potential hazards other than agent toxicity that can contribute to processing risk include the toxic effects of products of incomplete combustion of agent and other hazardous materials used in the disposal process, as well as the effects of fire or explosion. (Because the quantity of the products of incomplete combustion is substantially smaller than the original quantity of agent, combustion products generally represent a lesser hazard.) Release of toxic by-products can occur during process upsets, a possibility allowed for in an

---

NOTE: This appendix is adapted from NRC (1997).

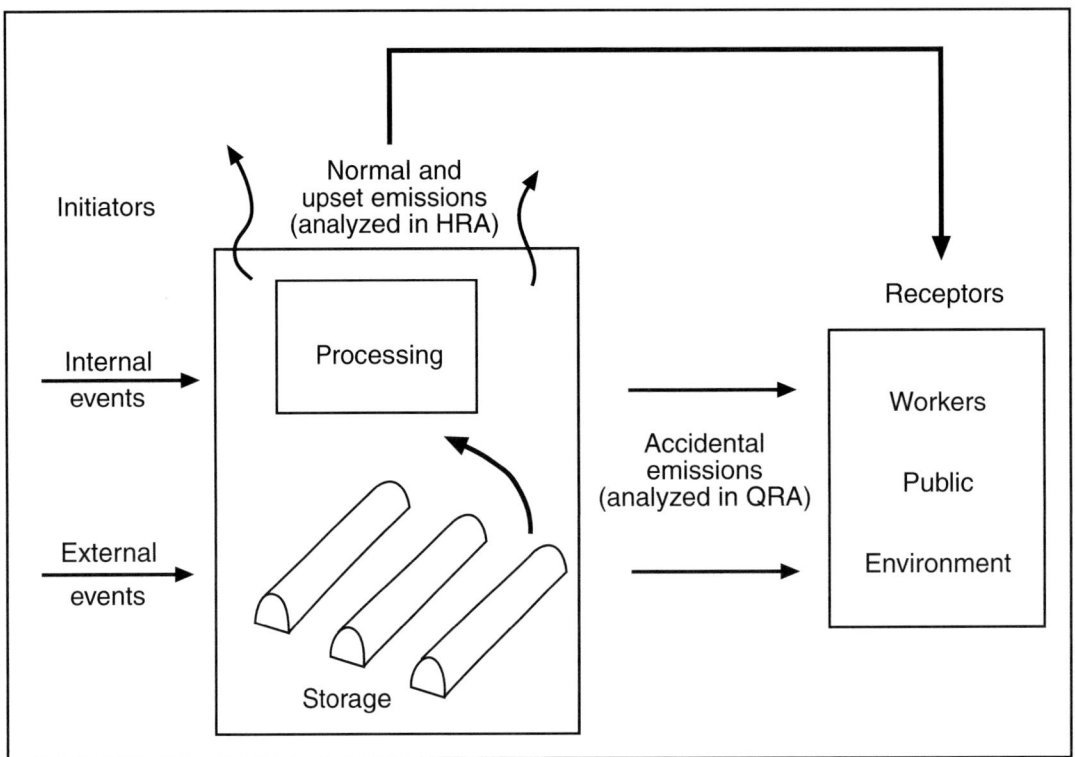

FIGURE E-1 Schematic illustration of risk elements at a chemical agent and munitions storage and destruction site. SOURCE: Adapted from NRC (1997).

upper-bound calculation in the QRA. External events such as earthquakes could cause the release of dangerous materials, such as propane or sodium hydroxide, from on-site storage tanks, as well as shutdown of the disposal process and possible one-time release of products of incomplete combustion from a furnace at shutdown.

Risk Receptors

There are three potential risk receptors: workers, the public, and the environment. Because of their proximity to the stockpile and agent-processing operations, *workers* are at risk from exposure to the acute lethal (and nonlethal) hazards associated with agent releases, regardless of the initiating event. They are also potentially at risk from long-term exposure to agent at very low concentrations and to the products and by-products of agent destruction. The only known latent effects are cancers following exposure to mustard (NRC, 2001a).[1]

---

[1] Workers are also susceptible to injury from ordinary industrial accidents (e.g., falls, burns, eye injuries, overheating in protective clothing), but such risks, which are not included in the Army's QRAs, can be better understood through safety inspections and analyses of injury rates and can be managed by adhering to safety practices proven in many industries.

Risks to the *public* stem primarily from releases of agent caused by external events, although the public could also be put at risk by long-term exposure to the products and by-products of agent destruction, if they were released into the environment as a result of chemical agent destruction processes. *Environmental* risk is associated almost exclusively with the release of agent and the products and by-products of agent destruction beyond site engineering controls.

Risk Measures

For humans (both workers and the public), the three potential consequences of the risk posed by either stockpile storage or agent destruction are acute lethality, acute and latent noncancerous health effects, and latent cancer. The potential adverse consequences for the environment are the contamination of land and water, and adverse effects suffered by native or endangered species.

Risk Mitigation

Risk is most effectively mitigated or prevented before a hazardous material is released. However, after a hazardous material has been released, but before it reaches a receptor, the consequences of the release can be reduced. Risk mitigation can include measures taken at the spill site (e.g., containing the spill) or at the receptor site (e.g., using protective masks), and emergency response measures (e.g., sheltering, evacuation). The Army's 1996 site-specific QRA takes into account some of these measures (U.S. Army, 1996a). However, the primary purpose of the QRA is to calculate a realistic estimate of risk to the public. The analysis is not structured to measure the effectiveness of the local Chemical Stockpile Emergency Preparedness Program (CSEPP).

Uncertainty

To provide the most realistic representation of risk, all forms of uncertainty are considered. Rather than assuming the existence of some representative condition prior to the accident scenario, a study models the full range of conditions and other uncertainties that can affect the scenario. Results include uncertainties in the frequency and consequences of each scenario. The upper uncertainty bound shown for the QRA risk estimates is a measure of the analysts' confidence in the results. There is a 95 percent chance that the risk is less than the upper bound.

For those readers desiring more details on risk assessment and risk management, Appendix A of the National Research Council (NRC) report *Risk Assessment and Management at Deseret Chemical Depot and the Tooele Chemical Agent Disposal Facility* (NRC, 1997) develops the bases for the presentations of risk (risk profiles and expected fatalities), explains how to interpret the results, and discusses various measures according to which risks can be compared.

QRA RESULTS: THE ARMY'S 1996 ANALYSIS

Results from the published QRA for the Tooele Chemical Agent Disposal Facility (TOCDF) (U.S. Army, 1996a) are used below to illustrate the form of QRA results.

Risk to the Public from Stockpile Storage

The risk-dominant initiating event for release of agent stored in the chemical stockpile at DCD is an earthquake (U.S. Army, 1996a). Although earthquakes are infrequent, they have widespread effects and could cause the release of much more chemical agent than other types of accidents. Seismic events that would contribute to storage risk are those with mean accelerations above 0.2 g and recurrence intervals of 1,000 years or more. Such earthquakes significantly exceed normal building code design values and thus can lead to failures of equipment and structures. Overall, according to the Army's 1996 QRA for TOCDF, earthquake-initiated events account for 82 percent of the average public fatality risk associated with continued storage of the stockpile; of the remaining 18 percent of the average public fatality risk, leaks of agent GB from ton containers account for 11 percent (Figure E-2).

An aircraft crash into storage structures and the electromagnetic effects of lightning (which could cause a fire in a storage igloo or cause an M55 rocket to ignite) were also considered in the Army's 1996 QRA. The results (see Figure E-2) indicated that the impact of these initiators is only 2 percent and 4 percent, respectively, of the total storage risk. Risks from normal stockpile maintenance, such as isolating leaking munitions, account for about 1 percent of the storage risk. These maintenance activities are infrequent, and the potential for a significant release is small because the number of munitions handled at any given time is limited.

Risk to the Public from Disposal Operations

The risk to the public from processing of chemical weapons is compared with the risk of continued storage in Figure E-3. The 1996 QRA put the risk level for the first campaign of GB disposal at about 0.00006 fatalities per year with a

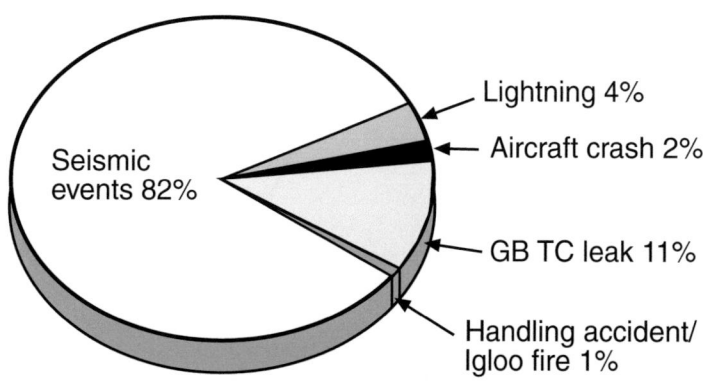

FIGURE E-2 Contributors to the average public fatality risk from continued storage at Deseret Chemical Depot. SOURCE: Adapted from U.S. Army (1996a).

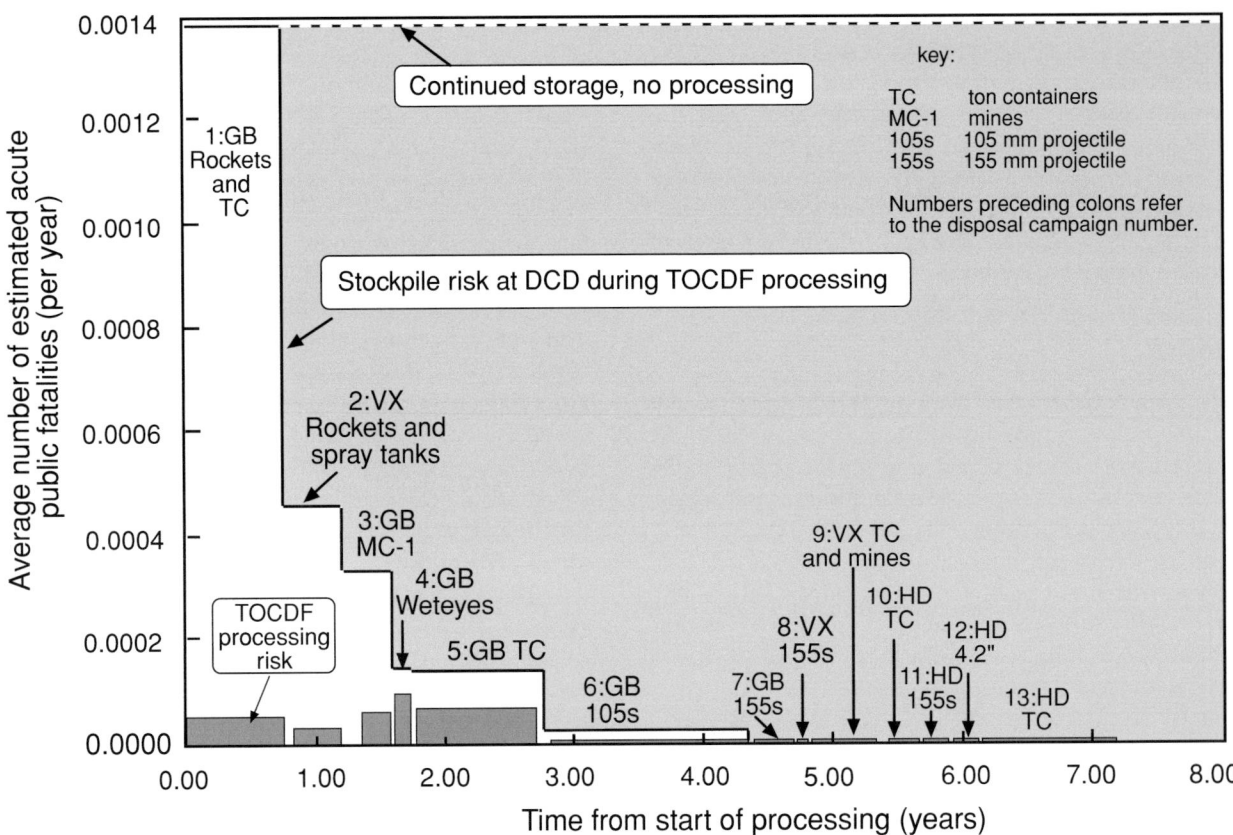

FIGURE E-3 Comparison of risks to the public during processing at Deseret Chemical Depot and the Tooele Chemical Agent Disposal Facility. All risks on these curves are shown on a per-year-of-operation basis so that they are directly comparable. The risks of continued storage, assuming no processing takes place, are indicated by the broken line. The vertical axis shows average public fatality risk per year, and the horizontal axis shows the time line for disposal. SOURCE: Adapted from U.S. Army (1996a).

processing duration of about 9 months (U.S. Army, 1996a). Note, however, that the stockpile storage risk decreases at the end of that time by two-thirds because the agent posing the greatest risk would be removed from the stockpile during the first disposal campaign.

By the end of the fifth campaign at TOCDF (GB ton containers nearly 3 years into disposal operations), the risk of both storage and processing have essentially disappeared. Nevertheless, although the risk is small, it is clear that storage risk is still much greater than processing risk and that accepting the processing risk for 3 years dramatically reduces the total risk.

Using the information shown in Figure E-3, risk managers at TOCDF ascertained the relative effects of various agent destruction campaigns. This information was used to reorder the disposal campaigns to minimize the total overall risk.

For disposal processing at TOCDF, the 1996 QRA results show that the risk of public fatalities is dominated by earthquakes (97.4 percent) as the most dangerous risk-initiating event (Figure E-4). A structural failure in the unpack area of the container handling building area caused by an

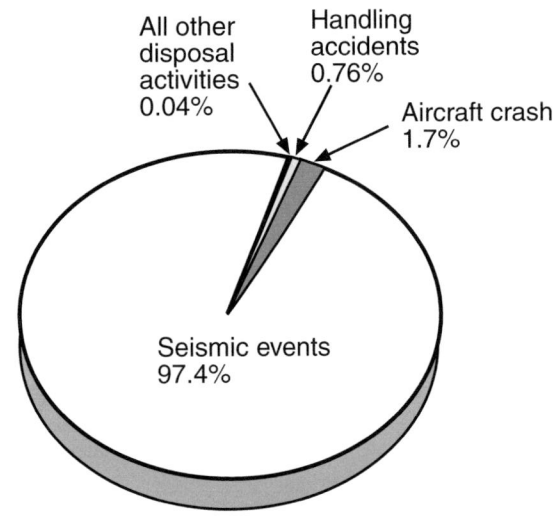

FIGURE E-4 Contributors to the average public fatality risk from disposal operations at DCD and TOCDF. SOURCE: Adapted from U.S. Army (1996a).

earthquake stronger than the building is designed to withstand could have severe consequences partly because munitions are unpacked in this area and are not protected by transport containers. The same earthquake would undoubtedly lead to the release of much more agent from the DCD storage area next to TOCDF.

The QRA results shown in Figure E-4 also indicate that internal events associated with processing account for less than 1 percent (i.e., 0.8 percent) of the risk at TOCDF and that nearly all of this risk is associated with handling rather than with actual agent destruction. The 1996 study credits the low risk of processing to the safety and mitigation features of the baseline system and the limited quantities of agent available for release during processing (U.S. Army, 1996a).

Risk to Workers from Disposal Operations

Workers at TOCDF, including all support and administrative staff at the facility or in nearby buildings and munition handlers responsible for removing munitions from the stockpile and transporting them to the disposal facility, were included in the Army's 1996 risk assessment. Although the study addressed only worker risks associated with accidents involving release of agent, processing and handling workers can be directly affected by the blast of an explosion, for example, or by dispersal of agent from an accident, and both of these effects were included. (Industrial-type risks, e.g., being crushed by a lift-truck, were not considered.) The QRA results indicated a 1 in 7 probability of a worker fatality in the total disposal-related worker population in the 7.1 years of disposal processing. Figure E-5 shows the contributors to the average risk of fatality for disposal-related workers.

The 1996 QRA indicates that risks to disposal workers from agent-related accidents are substantially higher than the risks to the public, as would be expected because of the proximity of the workers to the chemical agent. Small releases that would not have an impact at a significant distance could still be lethal to workers in the immediate area. According to the QRA, for the workforce of about 500 workers at TOCDF, if the 0.13 expected fatalities per 7.1 years of operation are dominated by single-fatality accidents, then the individual disposal-worker risk at TOCDF is about 4×10^{-5} per year.

The risk for other on-site workers (outside the TOCDF and DCD storage area) is evaluated in the same manner as the risk to the public. The probability of one or more fatalities for other on-site workers during the 7.1 years of disposal processing is 5×10^{-4} (1 in 2,000). With about 100 workers in this category, and assuming that most accidents cause a single fatality, the individual annual risk is 1×10^{-6} (1 in 1 million per year) for other on-site workers.

QRA: RECENT ADVANCES

The recent, as yet not published, Army QRAs for the third-generation facilities for chemical demilitarization have extended the methodology for assessing the effects of lightning, tornados, fires, events of special interest at each site, and human actions (and errors), and for understanding worker risk. The analysis for lightning draws on recent advances in tracking the position and strength of lightning strikes throughout the United States and new data that corrects long-held assumptions about the distribution of lightning. New technology for understanding the behavior of lightning within igloos has also been important. The analysis of risk from fire has shifted from the use of a physics model and nuclear power plant fire data to a more data-driven analysis, with more applicable data from chemical process plants having a significant impact on the results. After the

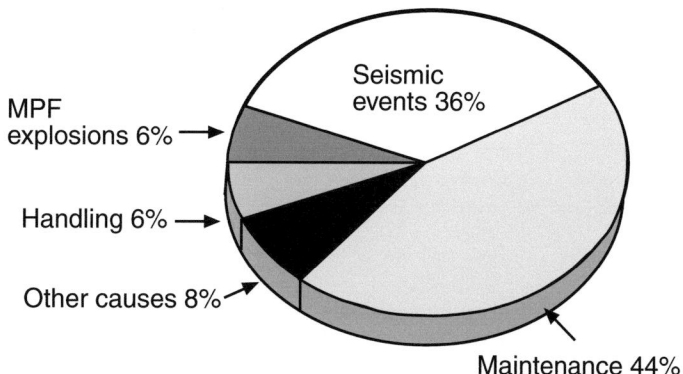

FIGURE E-5 Contributors to the average risk of fatality for disposal-related workers at DCD and TOCDF. SOURCE: Adapted from U.S. Army (1996a).

events of September 11, 2001, the Army decided to reconsider the publication of site-specific QRAs for the third-generation incinerator-based chemical demilitarization facilities. However, the committee ascertained that these QRAs confirm the dominance of the risk of continued storage of aging chemical weapons.

Improvements in the analysis of worker risk have resulted from an increased focus on worker activities and the adoption of more general methods for analyzing the effects of human error. For a number of reasons discussed in the NRC report *Risk Assessment and Management at Deseret Chemical Depot and the Tooele Chemical Agent Disposal Facility* (NRC, 1997), very little modeling of human performance was done in the 1996 QRA for TOCDF.[2] For example, workers directly involved in an accident were assumed to be killed, either from exposure to agent or from an explosion.

As attention in the chemical demilitarization program has shifted to include worker risk, more significant modeling of human action has been performed. None of these improved analyses have yet been published. A variety of human reliability analysis methods have been used (Gertman and Blackman, 1994). For ongoing work, new approaches that account for details of context and human cognitive function are being adapted (Hollnagel, 1998; USNRC, 2000). With more careful and complete analysis, new scenarios especially important to worker risk are being developed.

---

[2] Reporting the health effects for workers who are not directly involved, but who work in adjacent areas, would have been deceptive for several reasons:
- The TOCDF dispersion model may not properly capture the close-in dose.
- Projected latent effects from everyday activities (e.g., maintenance) are much greater than the latent effects from an agent accident and were not modeled in the QRA.
- The calculated latent risk to workers is very small compared with the acute risk.
- Worker risk from continued storage would have required assessing limited worker populations and restricted activity schedules that no longer existed at DCD.
- The primary goal of the QRA was to calculate the public risk from accidents in the operation of the TOCDF.

Appendix F

Causal Tree Analysis of December 3-5, 2000, Event at JACADS

A standard tool in reliability analysis, the causal tree or event tree is particularly useful in analyzing incidents to which operator actions contribute either positively or negatively. The committee recognizes that such trees are designed at the discretion of the analyst and should not be construed as reflecting scientific certainty. Therefore, Figure F-1, the causal tree for the December 3-5, 2000, event at JACADS, is provided for illustrative purposes. This analysis suggests that the incidents examined by the committee grew from normal activities into potentially dangerous events.

The activities charted can be categorized as ranging from normal operations through system response. In addition, some can extend back in time before the occurrence of the incident, e.g., latent failures.

- *Normal tasks*—tasks that the system was attempting to accomplish before the adverse event occurred. Examples are maintenance and operations.
- *Latent failures*—conditions present in the system for some time before the incident, but evident only when triggered by unusual states or events. Examples include equipment design deficiencies, unexpected configurations of munitions, or routine ignoring of standard operating procedures.
- *Active failures*—events *before* which there were no adverse consequences and *after* which there were. Active failures are usually the result of personnel decisions or actions. These same actions may have resulted in safe outcomes on previous occasions, but in the incidents examined by the committee, such actions combined with latent failures to cause some adverse consequences. Examples of active failures include use of the wrong procedure, incorrect performance of an appropriate procedure, or failure to correctly and rapidly diagnose a problem.
- *Immediate outcome*—the adverse state the system reached immediately after the active failure. Examples are release of agent, plant damage, or personal injury. Reporting and investigation flow charts supplied by the Army indicate that the severity of outcome often determines the incident's prominence for managers, the workforce, or the local community, which in turn drives subsequent responses. Incidents with more salient outcomes naturally receive more scrutiny, which may bias the data set used for analysis.
- *System responses*—actions taken to correct the effects and anticipate the aftereffects of an adverse outcome. Following each event, however, there is a system response that also needs to be analyzed. How did the system for incident response function? How did the management act to improve safety? Was an exposed worker properly treated? Were communities notified appropriately? How did the plant return to a normal state? How rapidly did it return? Finally, how was the system changed in light of the incident? This stage of analysis is considered in Chapter 4.

104

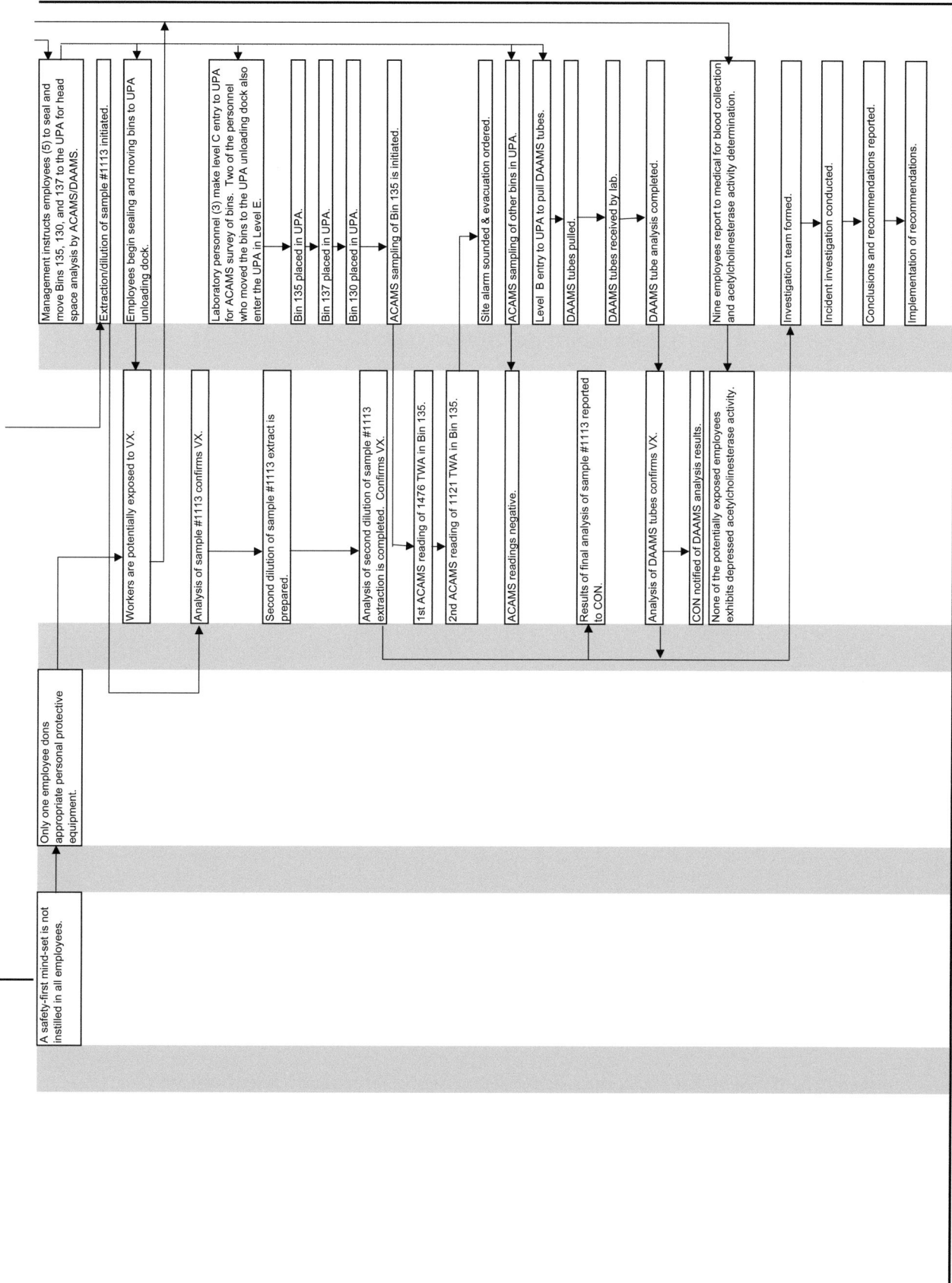

FIGURE F-1 Causal tree for December 3-5, 2000, JACADS event.

Appendix G

Memorandum of Understanding Between Deseret Chemical Depot and Tooele County for Information Exchange

TOOELE COUNTY CORPORATION
CONTRACT # 01-11-27

MEMORANDUM OF UNDERSTANDING

BETWEEN DESERET CHEMICAL DEPOT AND

TOOELE COUNTY

FOR INFORMATION EXCHANGE

1. **PURPOSE**

This Memorandum of Understanding (MOU) by and between Deseret Chemical Depot (DCD) and Tooele County outlines a cooperative, coordinated and pro-active information exchange process. It is designed by the parties to include planning considerations, hazard assessment and mitigation, coordinated emergency response, and effective protective actions. Its purpose is to achieve an enhanced state of emergency response readiness by providing the earliest possible hazard assessment and mitigation, and timely notification for protective action implementation, if required, for a chemical event or other type of emergency.

2. **OBJECTIVES**

The mutual objectives of the information exchange process outlined in this MOU are to: (1) Define specific chemical emergency categories; (2) Identify and display hazard predictions for chemical operations having the potential of producing chemical agent effects beyond the installation boundary; (3) Provide advance protective action recommendations for the potential emergency(s); and (4) Exercise daily activities that will mimic and reinforce emergency activities, thereby enhancing the notification and response capabilities of DCD and Tooele County.

To facilitate coordinated on-post and off-post emergency planning and response proficiency for a chemical event or other type of emergency, the parties agree to continuing information exchange,

---

NOTE: This reprinted appendix is the November 7, 2001, update of Utah DEQ (2000b).

using specified chemical event classification definitions; and following routine and emergency notification procedures as set forth herein.

3. COORDINATED DAILY INFORMATION EXCHANGE

DCD agrees to provide Tooele County with a work plan, prior to beginning daily chemical operations. The work plan will include an outline of the chemical tasks to be conducted, the estimated times that the operations will be in progress, agent and munition types, and current meteorological data. The work plan will specify tenant activities, even if there are no chemical operations scheduled. Hazard plots for every operation will be calculated and the maximum credible event (MCE) plot provided with a protective action recommendation (PAR) to Tooele County. DCD will provide updates to hazard plots if the wind shifts 30 degrees or more, or if the stability category changes.

Tooele County agrees to review DCD's daily chemical operations work plans, the hazard plots and the PARs from DCD at least twice daily. After review of the data provided by DCD, and giving consideration to off-post weather station readings and other community conditions, Tooele County will provide DCD and Utah Comprehensive Emergency Management with an appropriate protective action decision (PAD), including which communities would be instructed to evacuate or in-place shelter, specific evacuation routes, designated reception centers, etc. Tooele County will provide PAD updates as appropriate.

Major differences between PARs and PADs should be reviewed and discussed between Tooele County and DCD EOC personnel to achieve mutual understanding of the situation at hand, and to facilitate coordinated emergency response activities, when required. Consensus regarding the final PAD is preferred, where possible.

4. **EVENT CLASSIFICATION DEFINITIONS**

The following terms and emergency classification definitions are hereby agreed to by the parties, some of which are also detailed in the DCD Chemical Accident/Incident Response and Assistance Plan and the Tooele County Emergency Operations Plan, CSEPP Appendices:

(1) **Routine Leaker or Agent Detection Within Containment:** This classification will be used during munitions storage operations, such as when agent is detected in munitions within an unfiltered igloo or where there is no release to the atmosphere.

(2) **Category I: <u>Non-surety Emergency (Informational)</u>:** This level will be declared when events are likely to occur or have occurred that may be perceived as a chemical surety emergency or that may be of general public interest, but which pose no chemical surety hazard. These are further broken down into two groups as identified below:

(a) Non Chemical related non surety emergency – An occurrence such as a fire on the installation, transport of personnel by ambulance when lights and/or siren are utilized, loss of commercial power causing reduced operations or failure of electrical back-up sources, etc.

(b) Chemical related non surety emergency - Confirmed detection of chemical agent exceeding the established airborne exposure limits (AEL) outside of primary engineering controls but within secondary engineering controls, (not released to the atmosphere).

(3) **Category II (Site Response):** This category is broken down into two groups as listed below:

(a) <u>Limited Area Emergency</u>: This level will be declared when events are likely to occur or have occurred that involve agent release outside engineering controls, or approved chemical storage facilities, with the predicted chemical agent no-effects dosage distance not extending beyond the chemical limited area where the chemical event occurred.

(b) <u>Post Only Emergency</u>: This level will be declared when events are likely to occur or have occurred that involve agent release with the predicted chemical agent no-effects dosage distance extending beyond the chemical limited area, but not extending beyond the post-installation boundary.

(4) **Category III (External Response)**: <u>Community Emergency</u>: This level will be declared when events are likely to occur or have occurred that involve agent release with the predicted chemical agent no-effects dosage distance extending beyond the post-installation boundary.

5. NOTIFICATION PROCEDURES

DCD agrees that notification shall be made within 10 minutes when agent is detected in the atmosphere, i.e., outside engineering controls, and when other unusual circumstances occur. [Examples include agent detected outside an igloo or in an incinerator stack; personnel reported "down" or exhibiting symptoms of agent exposure; unscheduled shutdown of operations at Chemical Agent Munitions Disposal Systems (CAMDS) or Tooele Chemical Disposal Facility (TOCDF), transportation of agent to or from DCD, etc.]

Suspected but not confirmed agent releases outside of primary engineering controls, but within secondary engineering controls, need not be reported until agent presence is confirmed. However, DCD should provide a courtesy "heads up" call when agent readings are detected outside primary engineering controls and are awaiting confirmation or non-confirmation. Updated information will be provided to Tooele County upon notification of final disposition. Agent detection in storage igloos will be reported when initially identified. Upon placement of operational filters subsequent daily monitoring results need not be reported. Leaking munitions discovered during first entry monitoring/leaker isolation will be reported.

DCD agrees to use the dedicated "Chemical Notification Hotline" phone, as the primary means of notification for routine leakers and other occurrences of chemical agent detection outlined above, as well as for events falling into the defined chemical event classifications. The information provided by

DCD to Tooele County and Utah County shall follow the Chemical Notification Form. (See Attachment A.) A facsimile copy of the completed Chemical Notification Form will be provided by DCD as verification of data communicated verbally. A facsimile for a "heads up" call will be provided only after the final disposition of the agent detection is determined. DCD will provide Tooele County and Utah County updated information as soon as it becomes available.

For Category II, Limited Area, Post Only, and Category III, Community level emergencies, DCD will also provide, as soon as possible, an "EMIS Event Notification", as well as updated hazard plots and revised protective action recommendations as required. Until updated protective action recommendations are prepared and transmitted, the most recent work plan hazard plot and protective action recommendation will be implemented. Tooele County will provide notification to Utah Comprehensive Emergency Management according to established procedures.

As a result of an unusual occurrence, or if the potential exists for a chemical event, DCD agrees that "heads up" notification shall be made at the earliest possible opportunity, even if all the information needed to complete a Chemical Notification Form is not available. Frequent updates will be provided, communicating new information as soon as it becomes available.

Tooele County agrees to provide reciprocal information regarding the PAD made by Tooele County officials, including traffic access and control, evacuation routes, designated reception centers and other evacuee support, etc. Periodic updates will be provided, along with available information on any activities that could affect safe evacuation from the installation and surrounding off-post communities, such as train derailments, major traffic accidents, hazardous materials spills, etc.

6. TERMS

This agreement shall take effect upon signature and shall continue until amended or terminated, in writing, with thirty days notice.

DESERET CHEMICAL DEPOT:

_____ Date 28NOV01
PETER C. COOPER
Colonel, CM
Commanding

TOOELE COUNTY:

_____ Date 11/29/2001
DENNIS L. ROCKWELL
Tooele County Commissioner

_____ Date 11/20/01
DOUGLAS J. AHLSTROM
Attorney, Tooele County

ATTEST

_____ Date 11-20-01
DENNIS D. EWING
Clerk, Tooele County

CHEMICAL NOTIFICATION FORM

ACTUAL EVENT ☐   HEADS UP ☐   DRILL /EXERCISE ☐

For Use by County
Received by _____ Initial Report ☐   Update ☐

Received On: County Hotline ☐   Commercial Phone ☐   Other _____

PART I - REPORT SOURCE INFORMATION
References: (1) AR 50-6 (2) SB 742-1
1. Reported to County By: _____   Date _____   Time _____

PART II – EMERGENCY CLASSIFICATION LEVEL
2. Leaker Report ☐   CAT I   Non-Surety Emergency: Non Chemical ☐   Chemical ☐
                     CAT II: Limited Area Emergency ☐   Post-Only Emergency ☐
                     CAT III: Community Emergency ☐

PART III - ESSENTIAL ELEMENTS OF INFORMATION
3. Date Discovered _____ Time of Detection: _____
4. Installation: DCD ☐   TOCDF ☐   CAMDS ☐
   Igloo # _____   Station # _____   Station # _____
   Bldg # _____   Bldg # _____   Bldg # _____
                                            Igloo # _____

5. Munition Type _____ Number of Munitions _____
6. Agent: GB ☐   VX ☐   H ☐   L ☐   Type: Vapor ☐   Liquid ☐

7. How Event Occurred: Leak ☐   Spill ☐   Fire ☐   Explosion ☐
8. Quantity of Agent Spilled/Released: _____
9. Operation Being Performed: _____
10. Detectors Used to Find: _____   Readings: _____
    Detectors Used to Confirm: _____   Readings: _____
11. Patients/Injuries (Explain): _____
_____

12. Assistance Required (Explain): _____
13. Corrective Action Taken: _____
_____
_____
_____

14. Other: _____
_____
_____

PART IV - PROTECTIVE ACTION RECOMMENDATION (PAR)
(Wind Direction From _____ degrees)
15. PAR for populations/communities affected: _____
_____
_____
_____

APPENDIX G

PART V – NOTIFICATIONS BY DCD RISK MANAGEMENT (SAFETY)

A. ENVIRONMENTAL: 833-4434   Name:_____ Time:_____

B. SBCCOM: *

     Duty Hours: DSN   584-2933/2960
                  Com. (410) 436-2933/4482   Name:_____ Time:_____

     Non-Duty: DSN   584-2148
                  Com. (410) 436-2148   Name:_____ Time:_____

C. AOC: DSN   227-0218
           Com. (703) 697-0218   Name:_____ Time:_____

D. IF NON-DUTY HOURS OR NON-WORKDAY, FAX MESSAGE TO:

    1. AOC: (Primary)   Commercial only   (703) 693-6290

       DA : (Alternate) Commercial only   (703) 693-5570   Time:_____

    2. SBCCOM: Commercial only   (410) 436-4496   Time:_____

* SBCCOM will notify AMC

TOOELE COUNTY CORPORATION CONTRACT # 00-09-07   AMSSB-ODC-OI-E12

Appendix H

Biographical Sketches of Committee Members

Charles E. Kolb, *Chair*, graduated from the Massachusetts Institute of Technology with a B.S. in chemical physics and from Princeton University with an M.A. and a Ph.D. in physical chemistry. Dr. Kolb is president and chief executive officer of Aerodyne Research, Inc., in Billerica, Massachusetts. His principal research interests include atmospheric and environmental chemistry, combustion chemistry, materials chemistry, and the chemical physics of rocket and aircraft exhaust plumes. He has served on several National Aeronautics and Space Administration panels dealing with environmental issues as well as on six previous National Research Council (NRC) committees and boards dealing with atmospheric and environmental chemistry. Dr. Kolb also served on the NRC's Committee on the Review and Evaluation of the Army Chemical Stockpile Disposal Program (member, 1993-1998; vice chair, 1998-2000). He is a fellow of the American Physical Society, the American Geophysical Union, the American Association for the Advancement of Science, and the Optical Society of America.

Dennis C. Bley is president of Buttonwood Consulting, Inc., and a principal of The WreathWood Group, a joint venture company that supports multidisciplinary research in human reliability. Dr. Bley has a Ph.D. in nuclear engineering from the Massachusetts Institute of Technology and is a registered professional engineer in the state of California. He has more than 30 years of experience in nuclear and electrical engineering, reliability and availability analysis, plant and human modeling for risk assessment, Bayesian diagnostic system development, and technical management. He has served on a number of technical review panels for U.S. Nuclear Regulatory Commission and U.S. Department of Energy programs and is a frequent lecturer in short courses for universities, industries, and government agencies. He is active in many professional organizations and is on the board of directors of the International Association for Probabilistic Safety Assessment and Management. Dr. Bley has published extensively on subjects related to risk assessment. His current research interests include applying risk analysis to diverse technological systems, modeling uncertainties in risk analysis and risk management, technical risk communication, and human reliability analysis.

Colin G. Drury is University at Buffalo Distinguished Professor of Industrial Engineering, concentrating on the application of human factors techniques to manufacturing and maintenance processes. After earning a Ph.D. from the University of Birmingham, United Kingdom, he was manager of ergonomics at Pilkington Glass. He has extensive publications on topics in industrial process control, quality control, aviation maintenance, and safety and is the North American editor of *Applied Ergonomics*. From 1988 to 1993, he was the founding executive director of the Center for Industrial Effectiveness. He is a fellow of the Institute of Industrial Engineers, the Ergonomics Society, and the Human Factors Ergonomics Society. Dr. Drury received the Bartlett Medal of the Ergonomics Society and the Fitts Award of the Human Factors Ergonomics Society, and he has served on a number of NRC committees. He was a member of the NRC's Committee on the Review and Evaluation of the Army Chemical Stockpile Disposal Program from 1991 to 1995.

Jerry Fitzgerald English is a partner in the law firm of Cooper, Rose and English, LLP, where she heads the Environmental Law Department. She received a B.A. from Stanford University and a J.D. from Boston College Law School and is a member of the bar for both New Jersey and the U.S. District Court for New Jersey. Within her practice, Ms. English assists companies and communities in their legal defense and clean-up of contaminated property. Additionally, she has served as a New Jersey state senator, counsel to the governor of New Jersey, a commissioner of the New Jersey Department of Environmental Protection, and a commissioner of the Port Authority of New York and New Jersey. Ms. English co-chairs the American Bar Association's Environmental Litigation Section on Superfund and Hazardous Waste and teaches the practical aspects of ongoing remediation at the New Jersey Institute of Technology. She is a fellow of the American Bar Foundation. Ms. English has published extensively on subjects related to environmental law. She has considerable experience in the legal aspects of hazardous waste and remediation.

J. Robert Gibson graduated from Mississippi State University with a Ph.D. in physiology. He is board-certified in toxicology by the American Board of Toxicology. Dr. Gibson retired from the DuPont Company at the end of 2001 and is now president of Gibson Consulting, LLC. He has more than 25 years of experience in toxicology and occupational safety and health and is an expert in plant safety. As a member of the NRC's Stockpile Committee he gained extensive experience with the Chemical Stockpile Disposal Program.

Hank C. Jenkins-Smith is a professor of public policy at the George H. W. Bush School of Government and Public Service at Texas A&M University in College Station, Texas. He holds the Joe R. and Teresa Lozano Long Chair of Business and Government at the Bush School. He was previously a professor of political science and director of the Institute for Public Policy at the University of New Mexico. Professor Jenkins-Smith's areas of research include science and technology policy, environmental policy, public perceptions of environmental and technical risks, and national security policy. Professor Jenkins-Smith has written books on the public policy process and policy analysis, and has served on a number of committees for the National Research Council.

Walter G. May, NAE, has a B.S. in chemical engineering and an M.S. in chemistry from the University of Saskatchewan and an Sc.D. in chemical engineering from the Massachusetts Institute of Technology. He joined the faculty of the University of Saskatchewan as a professor of chemical engineering in 1943. In 1948, he began a distinguished career with Exxon Research and Engineering Company, where he was senior science advisor from 1976 to 1983. From 1983 until his retirement in 1991, he was a professor of chemical engineering at the University of Illinois, where he taught process design, thermodynamics, chemical reactor design, separation processes, and industrial chemistry and stoichiometry. Dr. May has published extensively, has served on the editorial boards of *Chemical Engineering Reviews* and *Chemical Engineering Progress*, and has obtained numerous patents in his field. He is a member of the National Academy of Engineering and is a fellow of the American Institute of Chemical Engineers. He has received special awards from the American Institute of Chemical Engineers and the American Society of Mechanical Engineers. He is also a registered professional engineer in the state of Illinois. Dr. May was a member of the National Research Council Committee on Alternative Chemical Demilitarization Technologies, the Stockpile Committee, and the Committee on Decontamination and Decommissioning of Uranium Enrichment Facilities.

Gregory J. McRae is the Bayer Professor of Chemical Engineering at MIT. His academic education includes a Ph.D. in engineering from the California Institute of Technology (Caltech). Dr. McRae currently teaches undergraduate and graduate level courses in process modeling, control, optimization, and computer-aided design. Another research focus is product and process design to avoid environmental problems and understanding the scientific aspects of problems involving pollutant transport and transformations in multimedia environments. His other interests include computational chemistry, process dynamics, turbulent fluid flow, computational fluid dynamics, reaction engineering, sensitivity/uncertainty analysis of complex systems, nonlinear parameter estimation, parallel computing, numerical analysis, and the design of cost-effective environmental controls. Professor McRae is the recipient of numerous awards and prizes for his research in environmental and computational science, including the Presidential Young Investigator Award, the George Tallman Ladd Research Prize, the Forefronts of Computational Science Award, and a AAAS Environmental Science Fellowship. He is a member of Sigma Xi, the American Chemical Society, and the American Institute of Chemical Engineers.

Irving F. Miller is director of Technical Consulting Services at BioTechPlex, a company that specializes in developing new devices and methods for the diagnosis and treatment of disease, to aid in drug discovery, and to support the health care industry. He has a B.Ch.E from New York University, an M.S. from Purdue University, and a Ph.D. from the University of Michigan, all in chemical engineering. Dr. Miller is a professional engineer in the state of New York and an educator. He was head of the departments of Chemical Engineering and Bioengineering, and associate vice chancellor for research at the University of Illinois at Chicago, where he is professor of bioengineering and chemical engineering, emeritus. He is also a former dean of engineering and a professor of biomedical engineering, emeritus, at the University of Akron. His professional life is an interesting blend of chemical and bioengineering that included research in blood substitutes, drug release, transport phenomena in the lungs, mechanisms of pulmonary mucociliary clearance, and pulmonary response to irritants. Dr. Miller is a member of several professional organizations, a fellow of the American Institute for Medical and Biological Engineering and of the American Institute of Chemical Engineers, and a senior member of the Biomedical Engineering Society.

Donald W. Murphy, NAE, has a B.S. in chemistry from Harvey Mudd College and a Ph.D. from Stanford University in inorganic chemistry. Dr. Murphy recently retired from Bell Laboratories, Lucent Technologies, where he was the director of the Applied Materials Research Department. He is currently a visiting researcher in the Chemistry Department of the University of California at Davis and an independent consultant. Dr. Murphy's research interests center on the synthesis of inorganic materials and on energy storage and conversion. A fellow of the American Association for the Advancement of Science and a member of the American Chemical Society, the American Physical Society, and the National Academy of Engineering, Dr. Murphy has also published widely in his field.

Alvin H. Mushkatel has a B.A. from Ohio State University and an M.S. and a Ph.D. from the University of Oregon, all in political science; he is a professor in the School of Planning and Landscape Architecture, Arizona State University. In addition, Dr. Mushkatel is the founder and former director of the Office of Hazards Studies at Arizona State University. His research interests include emergency management, natural and technological hazards policy, and fiscal impact analysis. He has been a member of the National Research Council's Committee on Decontamination and Decommissioning of Uranium Enrichment Facilities, the Panel on Review and Evaluation of Alternative Chemical Disposal Technologies, and the Stockpile Committee from 1992 to 1998. His most recent research is focused on the intergovernmental policy conflicts involving high-level nuclear waste disposal and the role of citizens in decision-making processes.

W. Leigh Short, with a Ph.D. in chemical engineering from the University of Michigan, retired as a principal and vice president of Woodward-Clyde responsible for the management and business development activities associated with the company's hazardous waste services in Wayne, New Jersey. Dr. Short has expertise in air pollution, chemical process engineering, hazardous waste services, feasibility studies and site remediation, and project management. He has taught courses in control technologies, both to graduate students and as a part of EPA's national training programs. He has served as chairman of the NO_x control technology review panel for the EPA.

Leo Weitzman received his Ph.D. in chemical engineering from Purdue University. He is a consultant with 29 years of experience in the development, design, permitting, and operation of equipment and facilities for treating hazardous waste and remediation debris. Dr. Weitzman has extensive experience in the disposal of hazardous waste and contaminated materials by thermal treatment, chemical reaction, solvent extraction, biological treatment, and stabilization. He has published more than 40 technical papers. Dr. Weitzman is currently serving on the National Research Council's Committee on Review and Evaluation of Alternative Technologies for Demilitarization of Assembled Chemical Weapons: Phase 2.

Appendix I

Committee Meetings

MAY 29-30, 2001—WASHINGTON, D.C.

Presenters

Chemical Weapons Demilitarization
Andrew Roach, *Staff Engineer, Operations Division*
Office of the Project Manager for Chemical
    Demilitarization

Congressional Perspective
David J. Cosco, *Staff*
Office of Congressman Bob Riley, AL

Previous Chemical Events
Gregory St. Pierre, *Chief, Risk Management & Quality
    Assurance*
Office of the Project Manager for Chemical
    Demilitarization

JUNE 28-29, 2001—JOHNSTON ATOLL, PACIFIC

Presenters

Orientation
Gary McCloskey, *JACADS Site Project Manager*

**Environment/Regulatory Requirements/Emergency
    Response**
Richard Mobley, *Emergency Preparedness Manager*
David Sewell, *Environmental Manager*

Systems
Paul Weader, *Engineering Manager*
Angelos Angeloupulos and Dave Ott, *Chief Systems
    Engineer*
Rick Mickey, *QA/QE Manager*
Louise Maxfield, *Configuration Manager*
Chuck Townsley, *Maintenance Manager*

Miscellaneous Activities
David Faucher, *CFS Manager*
Jerry Tiller, *Project Control Manager*

Operations and Safety
Dwayne Barks and Cindy Collins

Laboratory/Monitoring
Nollie Swynnerton, *Laboratory Manager*

Site Management
Steve Kirkendall, *Program Director*
Roger Dickerman, *Acting General Manager*
Heather Eusmurgle, *Acting Operations Manager*

Phil Carnes, *Staff Assistant*
Bill Lothinger, *Material Supply Manager*
D. Thomas, *Contract Manager*
Bill Knight, *Technical Supply Service Manager*

JULY 25-27, 2001—SALT LAKE CITY, UTAH

Presenters

Programmatic Lessons Learned
Rolf Dietrich, *PLL Task Manager*
Office of the Program Manager for Chemical Demilitarization

TOCDF Process and Operational Activities
James Colburn, *EG&G General Manager*

Overview of Chemical Events
Jim Hendricks, *TOCDF Site Project Manager*
James Colburn, *EG&G General Manager*

TOCDF Overview
Jim Hendricks, *TOCDF Site Project Manager*

TOCDF Safety and Risk Management
Tom Kurkjy, *EG&G Deputy General Manager*

Utah Citizens Advisory Commission
David Osler, *Member*

OCTOBER 18-19, 2001—WOODS HOLE, MASSACHUSETTS

Presenters

A Perspective of the TOCDF Process and Operational Activities
Gary Harris, *Former Employee, TOCDF*

Another Perspective of the TOCDF Process and Operational Activities
Suzanne Winters, *Chairman, Utah Citizens Advisory Commission*

NOVEMBER 12-13, 2001—WASHINGTON, D.C.

Meeting objectives: Consider and mark up first full message draft; discuss path to the coordination draft; conduct writing sessions.

No Presenters

DECEMBER 3-4, 2001—ANNISTON, ALABAMA

Presenters

Meet with Calhoun County Commissioners
Randy Wood, *Chairman*

How JACADS/TOCDF Lessons Learned Have Assisted ANCDF
Doug Hamrick, *Westinghouse Anniston Plant Manager*

Citizens Advisory Commission and General Public
Erme Wilkins, *Chairperson*

Calhoun County Emergency Operations Center
Mike Burney, *Director*
Emergency Operations

JANUARY 14-15, 2002—WASHINGTON, D.C.

Meeting objectives: Discuss project plan and updated draft report outline; discuss path to the coordination draft; conduct writing sessions.